INTRODUCTION TO PROBABILITY

John E. Freund
Arizona State University

DOVER PUB

This Dover edition, first published in 1993, is an unabridged and slightly corrected republication of the work first published by the Dickenson Publishing Company, Inc., Encino, California, in 1973.

Manufactured in the United States of America
Dover Publications, Inc., 31 East 2nd Street, Mineola, N.Y. 11501

Library of Congress Cataloging-in-Publication Data

Freund, John E.
Introduction to probability / John E. Freund.—Dover ed.
p. cm.
Originally published: Encino, Calif. : Dickenson Pub. Co., 1973.
Includes bibliographical references and index.
ISBN 0-486-67549-1
1. Probabilities. I. Title.
QA273.F84 1993
519.2—dc20 93-9358
CIP

CONTENTS

4

EVENTS, 86

5

RULES OF PROBABILITY, 111

6

CONDITIONAL PROBABILITIES, 134

7

PROBABILITY DISTRIBUTIONS, 167

PREFACE

The study of probability can be fun, it can be a challenge, and it can be relevant. This would not have been the case a few decades ago, at least, not for the beginner, but the increased interest in statistics and the teaching of set concepts in the lower grades has changed everything around.

Not too long ago, whatever probability was taught on the elementary level followed the material on combinatorial methods in textbooks on algebra, and it usually consisted of what might best be described as some exercises in "mental gymnastics." Of course, the mathematics requirements of many colleges are, or were, intended to expose the student to some kind of rigorous mental discipline, and the author once taught at a college which gave each student the option of taking a course in mathematics (algebra or trigonometry), a course in logic, or a course in Latin. The place of mathematics in general education, thus, evolved from the desire to expose the student to some "mathematical-type" thinking, but since this was seldom, if ever, accomplished by the standard-type course in algebra or trigonometry, special courses have been designed to provide a cultural, or historical, survey of mathematics. It is the author's hope that this book will provide another alternative: the study of probability. As this subject is presented here, it exposes the reader to some mental discipline, mathematical as well as philosophical, but what is more important, *it exposes the reader to the kind of mathematical thinking which not only applies to simple problems of everyday life, but which also provides the basis for modern scientific thought.* By not shying away from controversial issues (the *meaning* of "probability," for example), this study of probability provides the reader with some ideas about the strengths as well as the weaknesses of mathematics, and it will provide him, or her, with a sound foundation for future studies in the quantitative methods of business and all of the sciences.

The exercises in this book, and there are many, are designed to *reinforce* the methods and the ideas explained in the text, and the reader should benefit greatly from trying as many as possible. For the instructor, a manual with worked-out solutions is available; for the student, there is a workbook, which consists of an outline as well as sets of objective-type (true-or-false, multiple-choice, and fill-in) questions.

The author is greatly indebted to Dr. Al Romano for his many helpful comments on the various drafts of the manuscript, to Doug for his help with the proof reading, and to Ray for his many helpful comments and suggestions "from the student's point of view," as well as his help in working out the solutions to the exercises. Finally, the author is indebted to the editorial staff of the Dickenson Publishing Company for their courteous cooperation in the production of this book.

Scottsdale, Arizona JOHN E. FREUND

INTRODUCTION

"How to live with uncertainties" is a problem which is as old as mankind, but only in the last few hundred years have answers been sought through mathematics. Before that, everything relating to chance was looked upon as *divine intent,* or as the French writer Anatole France put it: "Chance is the pseudonym God uses when He does not want to sign His name." Thus, it was considered impious, or even sacreligious, to try to analyze the "mechanics" of the supernatural through mathematics; indeed, some of the mathematicians connected with the early study of probability theory were persecuted for this very reason.

There are other reasons, of course, for the slow development of the mathematics of probability. Whereas games of chance provided the impetus for the mathematical study of probability, fundamental issues are still obscured by the superstitions of gamblers, professional as well as amateur, who all too often trust their intuition or wishful thinking in the defiance of logic. And when it comes to logic and the way we think, it must be understood that many of the things which seem "obvious" today, were not so a few hundred years ago. Thus, it was not until the advent of scientific thought, with its emphasis on observation and experimentation, that it even occurred to anyone that probability theory might be used in the study of the laws of nature, or that it might be applied to the solution of simple problems of everyday life.

Strangely enough, chance events, due to their very nature, tend to support or affirm superstitions. The reason for this is that *highly unlikely things happen all the time,* and the persons to whom such "miracles" occur usually find it very difficult to accept the fact that it was "pure luck." What we mean is this: If a lottery offering one big prize sells 2,000,000 tickets, the odds against each ticket are enormous (that is, its chances of winning are extremely slim), and yet it is an *absolute certainty* that one of the 2,000,000 tickets will have to win. Often, persons holding such "lucky tickets" cannot resist reading all sorts of things into their good fortune, and are unwilling to accept it as "mere chance." Indeed, the expressions "pure luck" and "mere chance" themselves suggest that there may be other kinds of luck or chance—perhaps helped along by *Lady Luck,* the modern counterpart of the Roman goddess *Fortuna* pictured on the cover of this book as she appeared on an eighteenth-century English merchant token.

Another handicap is the persistent, though nearsighted, view that uncertainties are caused by ignorance, and hence that there would be no need for probabilities if we could only know all there is to know about a given situation. We have referred to this view as *nearsighted,* because it overlooks the important fact that *when uncertainties exist with regard to single events, these uncertainties are replaced with virtual certainties when the argument is applied to a great aggregate of similar events.* For instance, we cannot be certain whether Mr. Brown, who is a heavy smoker, will develop lung cancer, but we can be *virtually certain* that there will be more cases of lung cancer among ten thousand heavy cigarette smokers than among ten thousand persons who do not smoke. Logically, this need not be the case, but you can bet your last dollar that it will.

As should be apparent from this brief introduction, the study of probability is not all "hardhearted, cold, and impersonal mathematics." Although superstitions and biased preconceptions can lead to all sorts of complications, these non-mathematical considerations, taken in the proper perspective, may actually add to the challenge and the fascination of the subject matter of this book.

1

POSSIBILITIES

INTRODUCTION

Suppose someone asked us which is more likely, that the next birthday of a person chosen at random will fall on a Monday *or* that the birthdays of two persons chosen at random will fall on the same day of the week. Not giving the matter much thought, we may answer quickly that the second alternative is less likely to happen because—there are so many more possibilities. Well then, let us see how many possibilities there are. In the first case, there are seven possibilities, depending on whether the person's next birthday falls on a Monday, Tuesday, Wednesday, Thursday, Friday, Saturday, or Sunday, and of these only Monday represents a "success." Thus, in the first case *one-seventh* of the outcomes are favorable. In the second case there are, indeed, more possibilities: both persons might have their birthday on a Monday, the first person might have his birthday on a Wednesday and the second person on a Tuesday, the first person might have his birthday on a Friday and the second person on a Sunday, and so on. Carefully preparing a list, we would arrive at the result that there are altogether 49 possibilities, of which *seven* (Monday and Monday, Tuesday and Tuesday, ..., and Sunday and Sunday) provide a "success." Thus, in the second case seven of the 49 possibilities are favorable, which is again *one-seventh,* and we may well conclude from this that randomly choosing a person whose next birthday falls on a Monday is as likely as choosing two persons whose birthdays fall on the same day of the week. If the reader feels that this is a foregone conclusion, or that it was obvious from the start, suppose we substitute days of the year for days of the week. *Is it as likely to choose a person whose birthday is on a given day of the year, say, March 12th, as it is to choose two persons whose birthdays are on the same day of the year?* The answer is "Yes," and the argument is the same as before, but the result may not be quite so obvious.

Our example has shown how counting possibilities can go a long way toward evaluating chances. We must be careful, though, because we tacitly assumed in each case that the possibilities were all equally likely, and if this is not the case, we can run into all sorts of difficulties. Suppose, for instance, that in a baseball World Series (in which the winner is the first team to win four games) the National League champion leads the American

League champion 3 games to 2. *What are the chances that the American League team will nevertheless win the series?* Clearly, there are the following possibilities:

(1) The National League team will win the sixth game and the series.
(2) The American League team will win the sixth game and then the National League team will win the seventh game and the series.
(3) The American League team will win the sixth game and also the seventh game to win the series.

Thus, one of the three possibilities is favorable to the American League team, which is *one-third,* but can we conclude from this that the chances of the American League team winning the series equal, say, those of rolling a 1 or a 2 with a balanced die (so that two of the six faces are regarded as a "success")? The answer is "No"—even if the two teams are evenly matched—and this should be apparent from the fact that in the third case the American League team has to win two games to win the series, while in the first case the National League team has to win only one. All this goes to show that although counting possibilities may help, it does not necessarily provide the answers; very often, counting possibilities constitutes but the first step of a more detailed analysis.

In connection with the counting of possibilities there are essentially two kinds of problems, the first of which will be discussed in the section that follows, and the second in the section beginning on page 8. The first kind of problem is that of *listing everything that can happen in a given situation,* and even though this may give the impression of being straightforward, it is often easier said than done. It would be quite a job, for example, to list the *millions* of five-man committees which can be selected from among a company's 124 employees, and even when the number of possibilities is relatively small, it can be difficult to make sure that none have been left out. The second kind of problem is that of *determining how many different things can happen without actually compiling a list,* and it is here that we shall learn, for example, how to arrive at the result that 225,150,024 different five-man committees can be selected from among a company's 124 employees.

TREE DIAGRAMS

On page 3 we said that there are 49 ways in which the birthdays of two persons can fall on the various days of the week, pointing out at the time that we could arrive at this result by carefully preparing a list. Perhaps "systematically" would have been a better word than "carefully," for proceeding systematically we find that when the first person's birthday falls

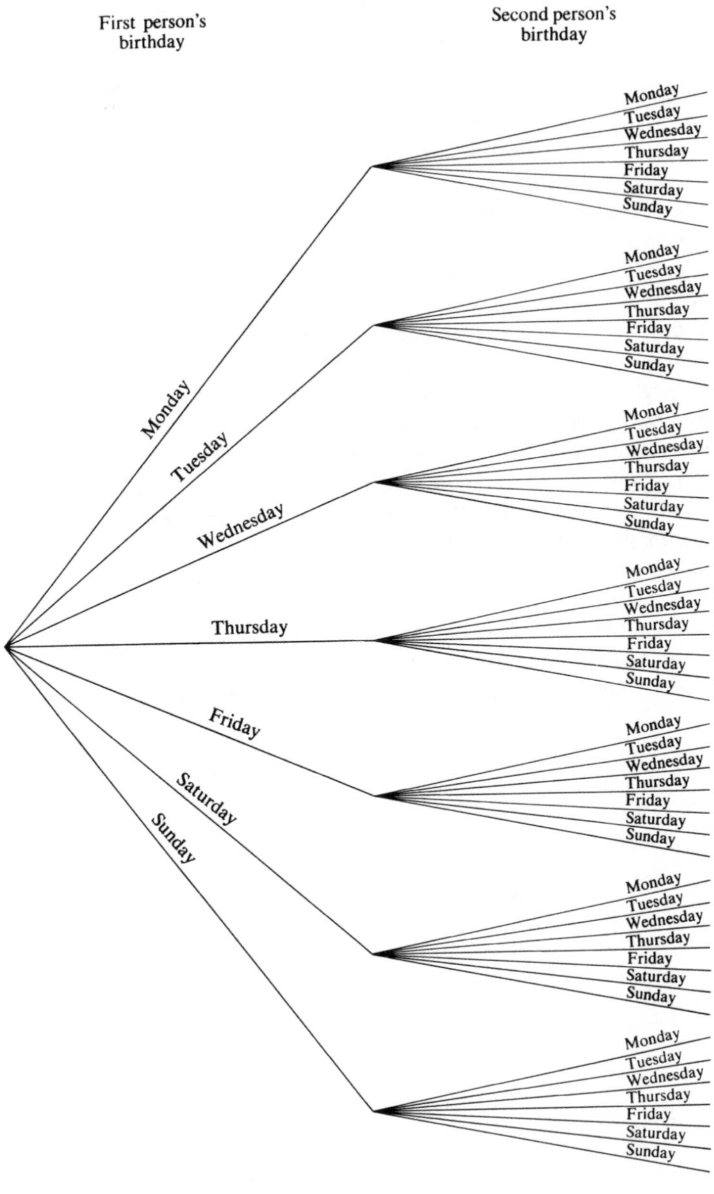

FIGURE 1.1 Tree diagram.

on a Monday there are seven possibilities corresponding to the second person's birthday falling on a Monday, Tuesday, Wednesday, Thursday, Friday, Saturday, or Sunday. Similarly, when the first person's birthday falls on a Tuesday there are the same seven possibilities corresponding to the second person's birthday falling on the different days of the week, and this argument applies also when the first person's birthday falls on a Wednesday, Thursday, Friday, Saturday, or Sunday. Thus, there are altogether *seven times seven,* or 49 possibilities.

In this kind of systematic listing of possibilities it often helps to refer to a diagram like that of Figure 1.1, which is called a **tree diagram**. This diagram shows that for the first person there are seven possibilities (branches) corresponding to his birthday falling on Monday, Tuesday, . . ., or Sunday. Then, for the second person there are seven branches emanating from each of the original branches again corresponding to the seven days of the week, and it can be seen that altogether there are 49 possibilities corresponding to the different paths along the "branches" of the tree. Starting at the top, the first path along the branches represents the case where the two birthdays fall on Monday and Monday, the second path represents the case where they fall on Monday and Tuesday, . . ., the eighteenth path represents the case where they fall on Wednesday and Thursday, ..., and the forty-ninth path represents the case where they fall on Sunday and Sunday.

For the other example of the preceding section, the one dealing with the World Series, we obtain the tree diagram of Figure 1.2. First, there are two possibilities corresponding to which team wins the sixth game. The first of

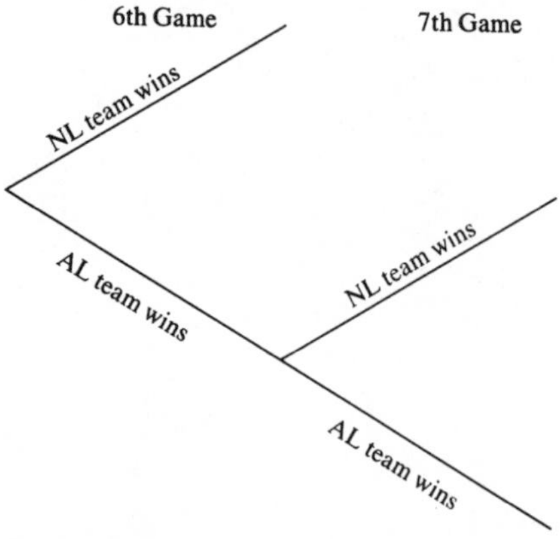

FIGURE 1.2 Tree diagram.

these does not figure in the next step since the National League team has already won, but emanating from the second branch there are again two branches corresponding to which team wins the seventh game. Clearly, there are three paths along the branches of this tree, and they represent the three possibilities listed on page 4. Of course, the original result was obtained very easily without the tree diagram, but this is not the case, for example, in Exercise 4 on page 11, where the reader will be asked to verify that there are *ten* possibilities (sequences of wins and losses for either team) after the National League team leads 2 games to 1 (instead of 3 games to 2).

To give another example in which the listing of possibilities is not straightforward and a tree diagram will help, suppose that a boy slips three pieces of candy in his pocket before going on a hike with his brother and a friend. What we are interested in is the number of different ways in which he can distribute *all* this "wealth." Clearly, there are many possibilities—he could keep all of the candy for himself, he could give two to his brother and one to the friend, he could keep one for himself and give one each to his brother and the friend, and so forth. Continuing this way, we may be able to complete the list, but unless we are very careful, the chances are that we will omit at least one of the possibilities.

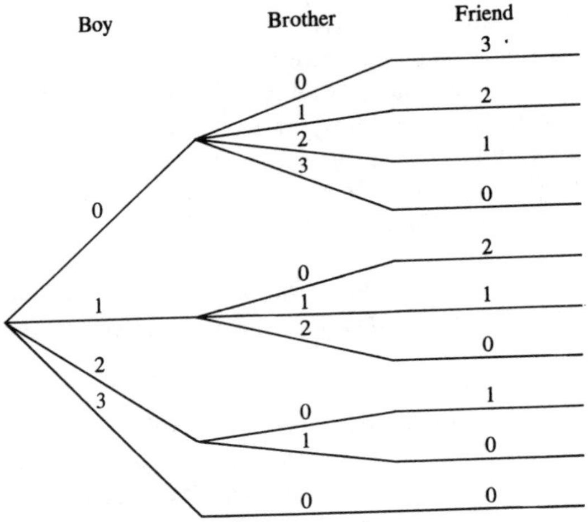

FIGURE 1.3 Tree diagram.

Referring to the tree diagram of Figure 1.3, we find that to begin with there are *four* possibilities (branches) corresponding to the boy's eating zero, one, two, or three pieces himself. Considering the brother next, we find that there are four branches emanating from the top branch, three from the second branch, two from the third branch, and one from the bottom branch. Evidently, there are again *four* possibilities (zero, one, two, or three) for the brother when the boy, himself, does not eat any of the candy, but only *three* possibilities (zero, one, or two) when the boy eats one piece of candy, *two* possibilities (zero or one) when the boy eats two pieces, and *one* possibility (zero) when the boy eats all of the candy. So far as the third step is concerned, there is only one possibility in each case, since the friend will get whatever candy, if any, remains.* Thus, it can be seen that there are altogether 10 possibilities corresponding to the different paths along the branches of the tree diagram of Figure 1.3—starting at the top, the first path corresponds to the boy's giving all of the candy to the friend, the second path corresponds to his giving one piece to his brother and two to the friend, . . ., and the tenth path corresponds to his eating all of the candy himself.

MULTIPLICATION RULES

The tree diagram of Figure 1.1 shows that there are *seven times seven* ways in which the birthdays of two persons can fall on the various days of the week. There are seven possibilities for the birthday of the first person, and for each of these there are seven possibilities for the birthday of the second person. Similarly, Figure 1.4 shows that if a restaurant offers pie, ice cream, cake, pudding, or fruit for dessert, which it serves with coffee, tea, or milk, there are *five times three* ways in which a person can order a dessert and a drink. These two examples illustrate the following rule, which is basic to most of the work of this chapter:

> **If a choice consists of two steps, of which one can be made in m ways and the other in n ways, then the whole choice can be made in $m \cdot n$ ways.**

This expresses what is sometimes called the "multiplication of choices," and to prove it, we have only to draw a tree diagram like those of Figures 1.1 and 1.4. First there are m branches corresponding to the possibilities in the first step, and then there are n branches emanating

*In Exercise 5 on page 11 the reader will be asked to show that when all, some, or none of the candy is distributed among the boy, his brother, and the friend, there are altogether 20 possibilities.

from each of these branches to represent the possibilities in the other step. This leads to $m \cdot n$ paths along the branches of the tree diagram, and hence $m \cdot n$ possibilities.

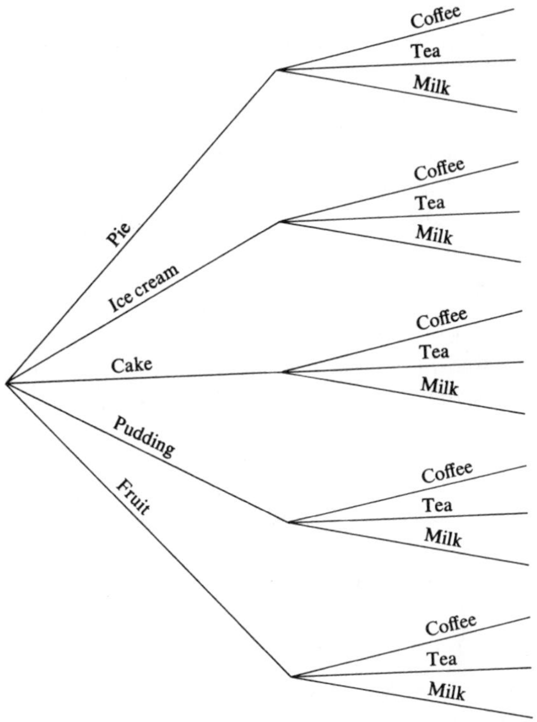

FIGURE 1.4 Tree diagram.

Thus, if there are 14 boys in a class and 11 girls, there are $14 \cdot 11 = 154$ ways in which an award can be given to one of the boys and one of the girls; if in a city election there are three candidates for mayor and two candidates for treasurer, there are $3 \cdot 2 = 6$ ways in which a person can vote for one candidate for each office; and if somebody wants to vacation in one of the eight Mountain states and travel by car, bus, train, or plane, he can plan his vacation in $8 \cdot 4 = 32$ different ways.

By using suitable tree diagrams, we can easily generalize the above rule for the "multiplication of choices" so that it applies to choices involving more than two steps. For k steps, where k is a positive integer, we have

If a choice consists of k steps, of which the first can be made in n_1 ways, the second in n_2 ways, ..., and the kth in n_k ways, then the whole choice can be made in $n_1 \cdot n_2 \cdot \ldots \cdot n_k$ ways.*

Thus, we simplify the number of ways in which each step can be made, and if a new-car buyer has the choice of four body styles, three different engines, and twelve colors, we find that he can make his choice in $4 \cdot 3 \cdot 12 = 144$ different ways. Furthermore, if he has the option of choosing a car with or without airconditioning, with or without an automatic transmission, and with or without power brakes, the total number of ways in which the new-car buyer can make his choice goes up to $4 \cdot 3 \cdot 12 \cdot 2 \cdot 2 \cdot 2 = 1{,}152$.

To consider another example, suppose that a test consists of 12 multiple-choice questions, with each permitting a choice of four alternatives. Applying the above rule with $n_1 = 4$, $n_2 = 4$, ..., and $n_{12} = 4$, we find that there are

$$4 \cdot 4 \cdot 4 \cdot 4 \cdot 4 \cdot 4 \cdot 4 \cdot 4 \cdot 4 \cdot 4 \cdot 4 \cdot 4 = 16{,}777{,}216$$

ways in which a student can check off his answers, and in only one of these cases will *all* his answers be correct. To make matters worse, in

$$3 \cdot 3 \cdot 3 \cdot 3 \cdot 3 \cdot 3 \cdot 3 \cdot 3 \cdot 3 \cdot 3 \cdot 3 \cdot 3 = 531{,}441$$

of the 16,777,216 possibilities *all* of his answers will be wrong.

EXERCISES

1. In a union election, Mr. Brown, Mr. Morris, and Mr. Mathews are running for president, while Mr. Adams, Mr. Mason, and Mr. Perkins are running for vice-president. Construct a tree diagram showing the nine possible outcomes of this election, and use it to determine the number of ways in which these union officials can be elected so that
(a) both of their names begin with the letter M;
(b) neither of their names begins with the letter M.
Can we conclude that it is *as likely* that both of their names will begin with the letter M as it is that neither of their names will begin with the letter M? Explain.

* In case the reader is not familiar with the use of **subscripts**, n_1 reads "n sub-one," n_2 reads "n sub-two," and so on. The advantage of this notation is that we can write the product of k numbers symbolically without knowing the value of k.

2. A businessman can go to work along four different routes, Routes A, B, C, or D, of which Route C is one-way, so that he cannot take it on the way home. Draw a tree diagram showing the various ways in which he can go to and from work, and use it to determine in what percentage of the possibilities he takes
(a) Route A to work;
(b) the same route both ways.
Can we conclude that he is just as likely to take *Route A* to work as he is to take the same route both ways? Explain.

3. In a traffic court, violators are classified according to whether or not they are properly licensed, whether their violations are major or minor, and whether or not they have committed any other violations in the preceeding 12 months.
(a) Construct a tree diagram showing the various ways in which this traffic court classifies violators.
(b) If there are 10 violators in each of the eight categories of part (a) and the judge gives each violator who is not properly licensed a stern lecture, how many of the violators will receive a stern lecture?
(c) If, furthermore, the judge gives a $50 fine to everybody who has committed a major violation and/or another violation in the preceding 12 months, how many of the violators will receive a $50 fine?
(d) How many of the violators will receive a stern lecture from the judge as well as a $50 fine?

4. Draw a tree diagram to show that there would have been ten possibilities in the example on page 4 if the National League team had led the American League team only by 2 games to 1. Since more than half of these possibilities require a seventh game, can we conclude that there is a better than fifty-fifty chance that there will be a seventh game? Explain.

5. With reference to the example on page 7, draw a tree diagram to show that there are altogether 20 ways in which the boy can distribute *all, some, or none* of the candy among himself, his brother, and the friend. Can we conclude from this that he is more likely to distribute two of the three pieces of candy as he is to distribute one or none? Explain.

6. To cut down on his television viewing, a child is allowed to watch *at most* three cartoon shows per day. Construct a tree diagram to determine the number of ways in which he can watch *at least* four cartoon shows on two consecutive days.

7. The pro at a golf course stocks two identical sets of women's clubs, reordering at the end of each day (for delivery early the next morning) if and only if he has sold them both. Construct a tree diagram to show that if he starts on a Monday with two sets of these clubs, there are altogether eight different ways in which he can make sales on the first two days of that week. In how many ways can he make sales on the first three days of that week?

8. A student can study either zero, one, or two hours for a history examination on any given night. Construct a tree diagram to show that there are ten different ways in which he can study altogether six hours for the examination on four consecutive nights.

9. If a college schedules four lecture sections and eight laboratory sections for Freshman Chemistry, in how many different ways can a student sign up for one of each? How many of these possibilities are left, if a student finds that two of the lecture sections and three of the laboratory sections are already filled?

10. A chain of furniture stores in California has four warehouses, of which two are in the Los Angeles area, and 15 retail outlets, of which six are in the Los Angeles area.

(a) In how many different ways can they ship an item from one of the warehouses to one of the stores?

(b) What percentage of the possibilities of part (a) involve only shipments from warehouses in the Los Angeles area to retail outlets in the Los Angeles area?

(c) What percentage of the possibilities of part (a) involve only shipments from warehouses in the Los Angeles area to retail outlets that are not in the Los Angeles area?

11. There are four different trails to the top of a certain mountain. In how many different ways can a person hike up and down the mountain if

(a) he must take the same trail both ways;

(b) he can, but need not, take the same trail both ways;

(c) he does not want to take the same trail both ways?

12. The five finalists in the Miss Arizona contest are Miss Cochise, Miss Yuma, Miss Pima, Miss Santa Cruz, and Miss Coconino.

(a) In how many different ways can the judges choose the winner and the first runner up?

(b) What percentage of the selections of part (a) will not include Miss Coconino?

(c) What percentage of the selections of part (a) will include neither Miss Yuma nor Miss Santa Cruz?

13. In a certain city, a child watching television on a Saturday morning has the choice between three different children's programs from 8 to 8:30, two different children's programs from 8:30 to 9, and three different children's programs from 9 to 9:30. In how many different ways can a child in this city plan his Saturday morning entertainment of watching children's programs (a) from 8:30 to 9:30, and (b) from 8 to 9:30?

14. Each Sunday, a newspaper carries a list of the 10 best-selling fiction books and the 10 best-selling non-fiction books.

(a) In how many different ways can we select one book from each list to take along on a trip?

(b) If, on a given Sunday, three of the fiction books and four of the non-fiction books are on the respective lists for the first time, in how many different ways can we select a book from each list that is *not* listed for the first time?

(c) With reference to part (b), in how many ways can we select four books from these lists so that one is a *fiction book that is listed for the first time,* one is a *fiction book that is not listed for the first time,* one is a *non-fiction book that is listed for the first time,* and one is a *non-fiction book that is not listed for the first time?*

15. In a drive-in restaurant, a customer can order a hamburger rare, medium rare, medium, medium well, or well done, and also with or without onions, with or without catsup, and with or without relish. In how many different ways can a person order a hamburger in this restaurant?

16. If a baseball squad has eight pitchers, 12 infielders, and nine outfielders, in how many different ways can a most-valuable-player award be given to one of the pitchers, one of the infielders, and one of the outfielders?

17. A travel bureau has arranged nine European tours for 1975 and eight for 1976. Find the number of ways in which one can choose

(a) one of these tours for each year.

If three of the 1975 tours and two of the 1976 tours include Amsterdam in the intinerary, find the number of ways in which one can choose

(b) one of the tours including Amsterdam for each year;

(c) one of the tours for each year *without visiting Amsterdam twice.*

18. A questionnaire sent through the mail as part of a market research study consists of eight questions, each of which can be answered in three different ways. In how many different ways can a person answer the eight questions on this questionnaire?

19. If a test consists of 15 true-false questions, in how many different ways can a student mark his test paper if
(a) he answers each question;
(b) he leaves any one of the questions unanswered?

20. How many different sums of money can be formed with one or more of the following coins: a half-dollar, a quarter, a dime, a nickel, and a penny? (*Hint*: there are two possibilities for each coin depending on whether or not it is included.)

21. Packing his belongings to go to college, a student has to decide what to do about his records. In how many different ways can he take along at least one of his 10 favorite records? (*Hint*: for each of the 10 records there are two possibilities, he can take it along or eave it.)

22. An art dealer has 10 large oil paintings and nine small oil paintings for sale. If his first customer buys one of the paintings and his second customer buys one of each size, how many different choices can the second customer make if
(a) the first customer buys one of the small paintings;
(b) the first customer buys one of the large paintings?

PERMUTATIONS

The multiplication rules of the preceding section are often used when several choices are made from one and the same set of objects, and the order in which these choices are made *is* of significance. This was the case, for example, in Exercise 12 on page 12, where the judges had to choose the winner and the runner-up in a beauty contest, and it would also be the case, say, if the 26 members of a club had to elect a president, a vice-president, a secretary, and a treasurer. The order *is* of significance, since the choice of Miss Yuma as Miss Arizona and Miss Coconino as runner-up differs from the choice of Miss Coconino as Miss Arizona and Miss Yuma as runner-up. Also, the election of Mr. Jones as president, Mrs. Murphy as vice-president, Mr. Brown as secretary, and Miss Smith as treasurer, and the election of Miss Smith as president, Mr. Brown as vice-president, Mrs. Murphy as secretary, and Mr. Jones as treasurer, are two different ways in which the four offices can be filled.

In general, if r objects are selected from a set of n objects, any particular arrangement of these r objects (say, in a list) is called a **permutation**. Thus, in the Miss Arizona example the winner and the runner-up can be, respectively,

Miss Cochise and Miss Yuma	*Miss Yuma and Miss Cochise*
Miss Cochise and Miss Pima	*Miss Pima and Miss Cochise*
Miss Cochise and Miss Santa Cruz	*Miss Santa Cruz and Miss Cochise*
Miss Cochise and Miss Coconino	*Miss Coconino and Miss Cochise*
Miss Yuma and Miss Pima	*Miss Pima and Miss Yuma*
Miss Yuma and Miss Santa Cruz	*Miss Santa Cruz and Miss Yuma*
Miss Yuma and Miss Coconino	*Miss Coconino and Miss Yuma*
Miss Pima and Miss Santa Cruz	*Miss Santa Cruz and Miss Pima*
Miss Pima and Miss Coconino	*Miss Coconino and Miss Pima*
Miss Santa Cruz and Miss Coconino	*Miss Coconino and Miss Santa Cruz*

and we refer to each of these choices as a permutation of two of the five finalists in the beauty pageant. Similarly,

$$3, 4, 5 \qquad 5, 4, 2 \qquad 2, 1, 4 \qquad 4, 2, 5$$

are four different permutations of three of the integers 1, 2, 3, 4, and 5, and

Tigers, Indians, Red Sox, Yankees

and

Orioles, Yankees, Senators, Tigers

are two different permutations of four of the six baseball teams in the Eastern division of the American League.

Of course, in each of the last two examples we listed only a few of the many possible permutations, and to verify that there are, in fact, 60 different permutations of three of the first five positive integers and 360 different permutations of four of the six baseball teams, we have only to refer to the following rule:

The total number of permutations of r objects selected from a set of n objects is $n(n-1)(n-2)\cdot\ldots\cdot(n-r+1)$.

To show this, we have only to refer to the multiplication rule on page 10 and observe that the first choice is made from the whole set of n objects, the second choice is made from the $n-1$ objects which remain after the first choice has been made, the third choice is made from the $n-2$ objects which remain after the first two choices have been made, ..., and the rth and final choice is made from the

$$n-(r-1) = n-r+1$$

objects which remain after the first $r-1$ choices have been made. Thus, the total number of permutations of r objects selected from a set of n objects, sometimes denoted ${}_nP_r$ or $P(n,r)$, is given by the product of r numbers, of which the first is n and subsequent numbers keep decreasing by one.

If we apply this rule to the two examples on page 15 which dealt with the selection of three of the first five positive integers and four of six baseball teams, we find that there are, indeed, $5\cdot4\cdot3 = 60$ permutations of three of the first five positive integers, and $6\cdot5\cdot4\cdot3 = 360$ permutations of four of the six baseball teams. Also, the 26 members of the club mentioned on page 14 can elect a president, a vice-president, a secretary, and a treasurer in $26\cdot25\cdot24\cdot23 = 358{,}800$ different ways, and also if somebody owns 38 tapes of popular music, he can listen to three of them in $38\cdot37\cdot36 = 50{,}616$ different ways (provided, of course, it matters in which order they are played).

If the judges in the Miss Arizona contest had to choose also a second, third, and fourth runner-up, they could have done so in $5\cdot4\cdot3\cdot2\cdot1 = 120$ different ways, and if we had asked for the total number of permutations of all six of the baseball teams (corresponding, say, to their standings at the end of the season), the answer would have been $6\cdot5\cdot4\cdot3\cdot2\cdot1 = 720$. This illustrates the fact that in the special case where $r = n$ (that is, when we are considering permutations of *all* the objects) the last factor in the formula for the total number of permutations is $n-r+1 = n-n+1 = 1$, so that

The total number of permutations of n objects taken all together is $n(n-1)(n-2)\cdot\ldots\cdot3\cdot2\cdot1$.

Thus, a person traveling through the East can visit Boston, New York, Washington, and Philadelphia in some order in $4\cdot3\cdot2\cdot1 = 24$ different ways, and a taxicab company can assign its eight drivers to its eight cabs in $8\cdot7\cdot6\cdot5\cdot4\cdot3\cdot2\cdot1 = 40{,}320$ different ways.

FACTORIALS

Since products of consecutive integers arise in many problems involving permutations and other kinds of special arrangements, we can often simplify matters by using the **factorial notation** in which $1! = 1$, $2! = 2\cdot1 = 2$, $3! = 3\cdot2\cdot1 = 6$, $4! = 4\cdot3\cdot2\cdot1 = 24$, $5! = 5\cdot4\cdot3\cdot2\cdot1 = 120, \ldots$, and in general $n!$, which reads "n factorial" or "factorial n," denotes the product

$$n(n-1)(n-2)\cdot\ldots\cdot3\cdot2\cdot1$$

for any positive integer n. To make certain formulas valid in all cases (as we shall see on page 18), it is customary also to let $0! = 1$ *by definition.*

The factorial notation can often be used to advantage when dealing with permutations, since any product of consecutive positive integers can be written as a quotient of two factorials. For instance, for $n = 8$ and $r = 4$ the total number of permutations is given by

$$8 \cdot 7 \cdot 6 \cdot 5 = \frac{8 \cdot 7 \cdot 6 \cdot 5 \cdot 4!}{4!} = \frac{8!}{4!}$$

and for $n = 16$ and $r = 5$ the total number of permutations is given by

$$16 \cdot 15 \cdot 14 \cdot 13 \cdot 12 = \frac{16 \cdot 15 \cdot 14 \cdot 13 \cdot 12 \cdot 11!}{11!} = \frac{16!}{11!}$$

In both of these examples, we, so to speak, filled in the missing factors by multiplying and dividing by a suitable factorial, and we also made use of the fact that in general $n!$ equals $n \cdot (n-1)!$, and also $n(n-1) \cdot (n-2)!$, $n(n-1)(n-2) \cdot (n-3)!$, $n(n-1)(n-2)(n-3) \cdot (n-4)!$, and so on.

The formula for the number of permutations of r objects chosen from a set of n objects can thus be written as

$$\begin{aligned} {}_nP_r &= n(n-1)(n-2) \cdot \ldots \cdot (n-r+1) \\ &= \frac{n(n-1)(n-2) \cdot \ldots \cdot (n-r+1) \cdot (n-r)!}{(n-r)!} \\ &= \frac{n!}{(n-r)!} \end{aligned}$$

This expresses the result in a more "compact" form, but it will, or may, simplify the calculations only if we can refer to a table of factorials, such as Table I at the end of this book. For instance, the total number of permutations of nine objects chosen from a set of 12 objects is given by the product $12 \cdot 11 \cdot 10 \cdot 9 \cdot 8 \cdot 7 \cdot 6 \cdot 5 \cdot 4$, which equals 79,833,600, but instead of actually multiplying all these numbers we can substitute $r = 9$ and $n = 12$ into $\frac{n!}{(n-r)!}$, look up the necessary factorials in Table I, and write

$$\frac{12!}{(12-9)!} = \frac{12!}{3!} = \frac{479{,}001{,}600}{6} = 79{,}833{,}600$$

To show that this method will not always provide simplifications (indeed, may be more cumbersome), let us find the number of permutations of two objects chosen from a set of 14 objects. According to the rule on page 8 the answer is $14 \cdot 13 = 182$, and it is much easier in this case to multiply the two numbers directly than to evaluate $\frac{14!}{12!}$ by looking up 12! and 14! in Table I. Thus, *in actual practice we use whichever method promises to involve the least amount of work.*

Note also that if we substitute $r = n$ into the formula that gives the total number of permutations as $\frac{n!}{(n-r)!}$, we get $\frac{n!}{(n-n)!} = \frac{n!}{0!}$ for the total number of permutations of n objects taken all together. Since the answer should be $n(n-1)(n-2)\cdot \ldots \cdot 3\cdot 2\cdot 1$ (namely, $n!$) according to the rule on page 16, this explains why we made the definition that $0! = 1$. Another justification which is sometimes given for this definition is that the equation $n! = n\cdot(n-1)!$ holds for *all positive integers* (including $n = 1$) only if we let $0! = 1$.

INDISTINGUISHABLE OBJECTS

Throughout this discussion it has been assumed that the n objects from which we make a selection are all distinct, namely, that they are distinguishable in some fashion; when this is not the case, there are complications. Evidently, if we write the letter A on eight slips of paper and then draw three of them out of a hat one after the other, the only possible thing we can get is A, A, and A. Thus, there is only one permutation in this case even though $r = 3$ and $n = 8$.

To illustrate what happens when some, but not necessarily all, of the objects are indistinguishable, suppose we want to determine the number of different ways in which we can arrange the letters in the word "room." If we distinguish for the moment between the two o's by labeling them o_1 and o_2, there are indeed $4! = 24$ different permutations of the symbols r, o_1, o_2, and m. However, if we drop the subscripts, then ro_1mo_2 and ro_2mo_1, for example, become the same permutation *romo*, and since *each pair of permutations with subscripts yields but one permutation without subscripts,* the total number of permutations of the letters in the word "room" is $\frac{4!}{2} = \frac{24}{2} = 12$.

Similarly, if we refer to the e's in the word "perceive" as e_1, e_2, and e_3, there are $8! = 40{,}320$ permutations of the symbols p, e_1, r, c, e_2, i, v, e_3, but since the three e's can be arranged among themselves in $3! = 6$ different ways, each permutation *without subscripts* corresponds to six permutations *with subscripts,* and hence there are only $\frac{8!}{3!} = \frac{40{,}320}{6} = 6{,}720$ different permutations of the letters in the word "perceive." (Another way of handling the above problems will be given in Exercise 14 on page 21.)

Also, if we refer to the *e*'s and *d*'s in the word "needed" as e_1, e_2, e_3, d_1, and d_2, there are $6! = 720$ permutations of the symbols n, e_1, e_2, d_1, e_3, d_2, but since the three *e's* can be arranged among themselves in $3!$ ways and the two *d*'s can be arranged among themselves in $2! = 2$ ways, each permutation *without subscripts* corresponds to $3!{\cdot}2! = 6{\cdot}2 = 12$ permutations *with subscripts,* and there are only $\frac{6!}{3!{\cdot}2!} = \frac{720}{12} = 60$ different permutations of the letters in the word "needed."

Using the same sort of reasoning, we can state more generally that the total number of permutations of n objects of which r_1 are alike, r_2 others are alike, ... and r_k others are alike is

$$\frac{n!}{r_1!{\cdot}r_2!{\cdot}\ldots{\cdot}r_k!}$$

For instance, the number of distinct arrangements of the 10 letters in the word "statistics," where there are three *s*'s, and three *t*'s, and two *i*'s, is $\frac{10!}{3!{\cdot}3!{\cdot}2!} = 50{,}400$.

To give an example where we are dealing with objects other than letters, suppose that a television director has four different commercials, one of which is to be shown in each of the eight time slots allocated to commercials during a 90-minute program. The question is, in how many different ways can he arrange his schedule if each of the commercials is to be shown twice? Clearly, $n = 8$, $r_1 = 2$, $r_2 = 2$, $r_3 = 2$, and $r_4 = 2$, so that substitution into the above formula yields the answer that he can schedule the commercials in $\frac{8!}{2!{\cdot}2!{\cdot}2!{\cdot}2!} = 2{,}520$ different ways. Had the director wanted to show two particular commercials three times and the other two commercials only once, substitution into the formula yields the result that he could have scheduled them in $\frac{8!}{3!{\cdot}3!} = 1{,}120$ different ways. In spite of this last example, which was more practical, it may seem to the reader that the material of this section is not very significant, for who cares about the number of ways in which one can arrange, say, the letters in the word "Mississippi?" As we shall see near the end of Chapter 7, however, there *are* important applications of this theory.

EXERCISES

1. On each business trip, a salesman visits three of the eight major cities in his territory. In how many different ways can he schedule his route (that is, the cities and their order) for such a trip?

2. Among the 16 applicants for four different teaching positions in an elementary school, only ten have M.A. degrees.
(a) In how many ways can these positions be filled?
(b) In how many ways can these positions be filled with applicants having M.A. degrees?
(c) If one of the positions requires an M.A. degree while for the others it is optional, in how many ways can the four positions be filled?

3. In how many ways can six new accounts be distributed among nine advertising executives, if none of them is to receive more than one of the new accounts?

4. A student has eight different textbooks, of which five are on business subjects and the other three are on foreign languages.
(a) In how many different ways can he arrange these books on a shelf?
(b) In how many different ways can he arrange these books on a shelf so that all the business books will be together and all the foreign language books will be together?

5. In how many different ways can the manager of a baseball team arrange the batting order of the nine players in his starting line-up? In what proportion of these batting orders will the pitcher bat last?

6. In how many ways can the 24 members of a sorority elect a president, a vice-president, and a secretary, assuming that no member can hold more than one office? In what percentage of these possibilities will Mary Jones, one of the 24 members of this sorority, be elected to one of the three offices?

7. Four married couples have bought eight seats in a row for a football game.
(a) In how many different ways can they be seated?
(b) In how many ways can they be seated if each couple is to sit together with the husband to the left of his wife?
(c) In how many ways can they be seated if each couple is to sit together?
(d) In how many ways can they be seated if all the men are to sit together and all the women are to sit together?
(e) In how many ways can they be seated if none of the men are to sit together and none of the women are to sit together? (*Hint*: add the number of possibilities where the first seat on the left is occupied by

a man to the number of possibilities where it is occupied by a woman.)

8. In how many different ways can four persons sit down to play bridge if
(a) it matters who sits in which chair;
(b) it matters only who sits next to whom?
[*Hint*: in part (b) arbitrarily seat one of the four persons and then see in how many ways the others can be seated.]

9. Use the hint to part (b) of Exercise 8 to find a general formula for the number of permutations of n objects arranged in a circle. Then use it to determine
(a) the number of ways in which eight persons can form a circle for a folk dance;
(b) the number of ways in which eight differently colored beads can be strung on an elastic band to form a necklace.
[*Hint*: in part (b) take note of the fact that the necklace can be turned over.]

10. Find the number of permutations of the letters in (a) "ship," (b) "meter," (c) "receive," and (d) "reader."

11. Find the number of permutations of the letters in (a) "monopoly," (b) "little," (c) "esteemed," (d) "nineteen," and (e) "parallel."

12. In how many ways (according only to manufacture) can nine cars place in a stock-car race, if three of the cars are Fords, three are Chevrolets, two are Plymouths, and one is a Dodge?

13. With reference to the illustration on page 19, in how many different ways can the television director schedule the commercials during the 90-minute program if
(a) he has eight different commercials, each of which is to be shown once;
(b) he has two different commercials, each of which is to be shown four times;
(c) he has two different humorous commercials and one serious commercial, each of the humorous commercials to be shown twice, and the serious commercial to be shown four times?

14. The number of permutations of the letters in the word "room," which we found on page 18, can also be obtained by distributing the letters r, o, o, and m among four positions; that is, we decide where each letter, taken one at a time, is to go. Thus, distributing first the letter r and then the letter m, we find that there are four places where we can put the letter r and then three places where we can put the letter m. Since this automatically leaves two positions for the two o's, we conclude that there are altogether $4 \cdot 3 = 12$ ways in which the four

letters can be distributed among the four positions, namely, that there are 12 different permutations of the letters in "room." This agrees, of course, with the result obtained on page 18. Use this method to
(a) verify the result obtained on page 19 for the number of permutations of the letters in "perceive;"
(b) rework part (b) of Exercise 10;
(c) rework part (c) of Exercise 10;
(d) rework part (c) of Exercise 11.

15. Check each of the following to see whether it is true or false:
(a) $6! = 6 \cdot 5 \cdot 4!$
(b) $\frac{10!}{10 \cdot 9 \cdot 8} = 7;$
(c) $\frac{1}{2!} + \frac{1}{2!} = 1;$
(d) $10! + 3! = 13!.$

COMBINATIONS

If a housewife wants to buy a half pound each of three of the 12 cheeses sold by a supermarket, she can do so in 220 different ways, but this is *not* the number of permutations of three objects chosen from a set of 12 objects. If we actually cared about the order in which she makes her selection, the answer would be the number of permutations, namely, $12 \cdot 11 \cdot 10 = 1{,}320$, but each choice of three cheeses would then be counted $3! = 6$ times. If we are not interested in the order in which the three cheeses are selected, there are thus $\frac{1{,}320}{6} = 220$ ways in which the housewife can choose three of the 12 cheeses, and we refer to this figure as the number of **combinations** of three objects chosen from a set of 12 objects.

A combination is thus the same as a subset, and when we ask for the number of combinations of r objects chosen from a set of n objects, we are simply asking "How many different subsets of r objects can be chosen from a set of n objects?" (By **subset** we mean any part of a set, including as special cases the whole set, when $r = n$, and the empty set, when $r = 0$.) To obtain a formula which supplies the answer to this question, we have only to observe that any r objects can be arranged among themselves in $r!$ permutations, which count only as *one* combination. Hence, the $n(n-1)(n-2) \cdot \ldots \cdot (n-r+1)$ different permutations of r objects chosen from a set of n objects contain each combination $r!$ times, and it follows that to obtain the number of combinations we have only to divide the number of permutations by $r!$. This leads to the following rule:

The number of combinations of r objects chosen from a set of n objects is

$$\frac{n(n-1)(n-2)\cdot\ldots\cdot(n-r+1)}{r!}$$

for $r = 0, 1, 2, \ldots,$ or n.

Substituting $\frac{n!}{(n-r)!}$ for $n(n-1)(n-2)\cdot\ldots\cdot(n-r+1)$ in accordance with the rule on page 17, we can write the above formula also as

$$\frac{n!}{(n-r)!\cdot r!}$$

Symbolically, we shall write the number of combinations of r objects chosen from a set of n objects as $\binom{n}{r}$, and we shall refer to these quantities as **binomial coefficients** for reasons to be explained on page 26. [The following are two alternate symbols used sometimes to denote the number of combinations of r objects selected from a set of n objects: ${}_nC_r$ and $C(n, r)$.]

To consider a few applications, let us see in how many ways a committee of five can be selected from among the 80 employees of a company, and let us also find the number of ways in which a research worker can choose eight of the 12 largest cities in the United States to be included in a survey. In the first problem we find that there are

$$\binom{80}{5} = \frac{80\cdot79\cdot78\cdot77\cdot76}{5!} = \frac{\overset{2}{\not{80}}\cdot79\cdot78\cdot77\cdot76}{\not{5}\cdot\not{4}\cdot3\cdot\not{2}\cdot1} = 24{,}040{,}016$$

ways of choosing the committee, and in the second problem we find that the research worker can choose the cities in

$$\binom{12}{8} = \frac{12!}{4!\cdot8!} = \frac{479{,}001{,}600}{24\cdot40{,}320} = 495$$

different ways. Also, in the first problem we used the first form in which the formula for the number of combinations was given in the rule on this page, and in the second problem we used the second form together with Table I. Actually, it would have been just as easy (or perhaps even easier) to use the first form of the formula also in the second problem;

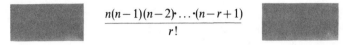

cancelling as many factors as possible before performing any multiplications, we would thus have obtained

$$\binom{12}{8} = \frac{12\cdot 11\cdot \overset{5}{\cancel{10}}\cdot 9\cdot \cancel{8}\cdot \cancel{7}\cdot \cancel{6}\cdot \cancel{5}}{\cancel{8}\cdot \cancel{7}\cdot \cancel{6}\cdot \cancel{5}\cdot \cancel{4}\cdot \cancel{3}\cdot \cancel{2}\cdot 1} = 11\cdot 5\cdot 9 = 495$$

To simplify the calculations even further in this last example, we might have made use of the fact that $\binom{12}{8} = \binom{12}{4}$ and in general

$$\binom{n}{r} = \binom{n}{n-r}$$

Clearly, when we choose r objects from a set of n objects we leave $n-r$ of the n objects, so that *there are as many ways of leaving (or choosing) $n-r$ objects as there are of choosing r objects.* [This rule can also be proved algebraically, as the reader will be asked to show in part (a) of Exercise 22 on page 32.] Making use of the rule in the second problem of the preceding paragraph, we can now write

$$\binom{12}{8} = \binom{12}{4} = \frac{12\cdot 11\cdot 10\cdot 9}{4!} = \frac{12\cdot 11\cdot \overset{5}{\cancel{10}}\cdot 9}{\cancel{4}\cdot \cancel{3}\cdot \cancel{2}\cdot 1} = 495$$

and the result is, of course, the same as before.

The rule which we have just proved plays an important role in the use of Table II at the end of this book, which enables us to read off the values of $\binom{n}{r}$ for $n = 0, 1, 2, \ldots,$ and 20. For instance, $\binom{12}{8}$ can be looked up directly—it is in the $\binom{n}{8}$ column in the row which corresponds to $n = 12$. On the other hand, $\binom{20}{14}$ cannot be looked up directly as there is no $\binom{n}{14}$ column in the table, but making use of the fact that $\binom{20}{14} = \binom{20}{6}$, we can find the answer in the $\binom{n}{6}$ column in the row which corresponds to $n = 20$. It is $\binom{20}{14} = \binom{20}{6} = 38{,}760$.

So far it has been assumed in this section that the n objects are all distinguishable, for otherwise there are complications worse than those on page 18. Suppose, for example, that we have two quarters, a dime, a nickle, and a penny, and we want to know how many different sums we can form with any two of these coins. Actually, there are very few possibilities in this case, but to proceed systematically, let us investigate separately the three cases where zero, one, and two of the coins we choose are quarters. Thus, we find that when *none* of the coins are quarters, there are $\binom{3}{2} = 3$ ways of choosing *two* of the other *three* coins; when *one* of the coins is a quarter, there are $\binom{3}{1} = 3$ ways of choosing *one* of the other *three* coins; and when *both* coins are quarters, there is only $\binom{3}{0} = 1$ way of choosing *none* of the other *three* coins. It follows that the answer is $3+3+1 = 7$.

Problems like this tend to get quite complicated and we shall not pursue them further in this book, but let us indicate briefly how the method of Exercise 14 on page 21 can be used to handle more complicated *permutation problems* involving indistinguishable objects. For instance, to find the total number of permutations of the letters in the word "needed," as on page 19, let us fill the six positions (which letter is to come first, which letter is to come second, and so on) by first choosing three positions for the *e*'s, then two positions for the *d*'s, and this will automatically leave only one position for the *n*. Thus, we find that there are $\binom{6}{3} = 20$ ways of placing the *e*'s, then $\binom{3}{2} = 3$ ways of placing the *d*'s, and hence altogether $20 \cdot 3 = 60$ permutations of the letters in the word "needed." Of course, this agrees with the result which we obtained on page 19.

BINOMIAL COEFFICIENTS

In elementary algebra we learn how to expand expressions such as $(a+b)^2$, $(a+b)^3$, $(a+b)^4$, ..., by actually multiplying them out term by term. For $(a+b)^3$, for example, we would thus get

$$
\begin{aligned}
(a+b)^3 &= (a+b)(a+b)(a+b) \\
&= a \cdot a \cdot a + a \cdot a \cdot b + a \cdot b \cdot a + a \cdot b \cdot b + b \cdot a \cdot a + b \cdot a \cdot b + b \cdot b \cdot a + b \cdot b \cdot b \\
&= a^3 + 3a^2b + 3ab^2 + b^3
\end{aligned}
$$

where each of the eight terms in the product of *three* letters, a or b, with one coming from each of the three factors $a+b$. For example, the three terms a^2b were obtained by multiplying the a of the first factor by the a of the second factor and the b of the third factor, by multiplying the a of the first factor by the b of the second factor and the a of the third factor, and by multiplying the b of the first factor by the a of the second factor and the a of the third factor. In other words, *there are three terms a^2b corresponding to the $\binom{3}{1} = 3$ ways in which we can choose one b and two a's, one from each factor* $a+b$. Similarly, there are *three* terms ab^2 corresponding to the $\binom{3}{2} = 3$ ways in which we can choose two b's and one a, one from each factor $a+b$; but there is only *one* term a^3 and *one* term b^3 corresponding to the $\binom{3}{0} = 1$ and $\binom{3}{3} = 1$, ways in which we can choose zero b's (namely, three a's) or three b's, one from each factor $a+b$. Thus, we could have written the expansion of $(a+b)^3$ as

$$(a+b)^3 = \binom{3}{0}a^3 + \binom{3}{1}a^2b + \binom{3}{2}ab^2 + \binom{3}{3}b^3$$

where $\binom{3}{0}$, $\binom{3}{1}$, $\binom{3}{2}$, and $\binom{3}{3}$, are, respectively, the *coefficients* of a^3, a^2b, ab^2, and b^3.

If we apply the same sort of reasoning to the expansion of $(a+b)^n$, where n is a positive integer, we obtain the following result, called the **binomial theorem**:

$$(a+b)^n = \binom{n}{0}a^n + \binom{n}{1}a^{n-1}b + \binom{n}{2}a^{n-2}b^2 + \ldots + \binom{n}{n-1}ab^{n-1} + \binom{n}{n}b^n$$

Thus, $\binom{n}{r}$ is the coefficient of $a^{n-r}b^r$ in the expansion of $(a+b)^n$, and this explains why we refer to the combinatorial symbols $\binom{n}{r}$ as *binomial coefficients.*

To illustrate the use of the binomial theorem as given above, let us find the expansions of $(a+b)^6$ and $(a+b)^{10}$, which would evidently require a lot of work if we had to multiply out term by term. In contrast, we have only to look up the values of $\binom{6}{0}$, $\binom{6}{1}$, $\ldots$, $\binom{6}{6}$, and $\binom{10}{0}$, $\binom{10}{1}, \ldots, \binom{10}{10}$ in Tablé II, and we can immediately write the results as

$$(a+b)^6 = a^6+6a^5b+15a^4b^2+20a^3b^3+15a^2b^4+6ab^5+b^6$$

and

$$(a+b)^{10} = a^{10}+10a^9b+45a^8b^2+120a^7b^3+210a^6b^4+252a^5b^5 \\ +210a^4b^6+120a^3b^7+45a^2b^8+10ab^9+b^{10}$$

There are several special relationships among binomial coefficients which serve to simplify calculations. For instance, we already learned that $\binom{n}{r} = \binom{n}{n-r}$ for $r = 0, 1, 2, \ldots,$ *or* n, and we saw that this not only helps to simplify calculations, but it comes in handy in connection with the use of Table II. The following is another important relationship among binomial coefficients, which holds for any positive integer n greater than 1 and for $r = 1, 2, \ldots,$ or $n-1$:

$$\binom{n}{r} = \binom{n-1}{r-1} + \binom{n-1}{r}$$

For instance, $\binom{8}{5} = \binom{7}{4}+\binom{7}{5}$ and $\binom{12}{7} = \binom{11}{6}+\binom{11}{7}$, as the reader will be asked to verify in parts (a) and (c) of Exercise 21 on page 31.

Leaving it to the reader to prove the above formula algebraically by expressing the three binomial coefficients in terms of factorials, which is really simpler, in part (b) of Exercise 22 on page 32, let us prove it here by arguing as follows: There are $\binom{n}{r}$ ways in which a set of r objects can be chosen from a set of n objects, but if we single out one of the n objects, this total should also be given by the *sum* of the number of ways in which we can choose r objects *including the one we have singled out* and the number of ways in which we can choose r objects *without including the one we have singled out*. In the first case, there are $\binom{n-1}{r-1}$ ways in which we can choose the other $r-1$ objects, in the second case there are $\binom{n-1}{r}$ ways of choosing r objects other than the one we have singled out, so that altogether there are $\binom{n-1}{r-1}+\binom{n-1}{r}$ ways of choosing r of the n objects. This completes the proof.

An immediate application of this rule is in the construction of what is commonly known as **Pascal's triangle**, which provides a quick-and-easy way of determining binomial coefficients. "Pascal's triangle" is the name given to the following array of numbers:

$$\begin{array}{ccccccccccc} & & & & 1 & & 1 & & & & \\ & & & 1 & & 2 & & 1 & & & \\ & & 1 & & 3 & & 3 & & 1 & & \\ & 1 & & 4 & & 6 & & 4 & & 1 & \\ 1 & & 5 & & 10 & & 10 & & 5 & & 1 \\ . & . & . & . & . & . & . & . & . & . & . \end{array}$$

where each row begins with a 1, ends with a 1, and each other entry is the *sum* of the nearest two entries in the row immediately above. The importance of this triangular array of numbers is that *the rth entry of the nth row is the binomial coefficient* $\binom{n}{r-1}$; for instance, the *third* entry of the *fourth* row is the binomial coefficient $\binom{4}{2}$, which equals 6, and the *fifth* entry of the *fifth* row is the binomial coefficient $\binom{5}{4}$, which equals 5. To prove this we have only to observe that $\binom{n}{0}$ and $\binom{n}{n}$, like the first and last entries of each row, are always equal to 1, and that in accordance with the formula $\binom{n}{r} = \binom{n-1}{r-1} + \binom{n-1}{r}$ which we just proved, each other entry *should* be the sum of the nearest two entries in the row immediately above.

In any case, the above array goes only as far as $n = 5$, but following the method of construction we find that the entries of the sixth row are, respectively, 1, $1+5 = 6$, $5+10 = 15$, $10+10 = 20$, $10+5 = 15$, $5+1 = 6$, and 1, and as can be seen from Table II, these are the binomial coefficients $\binom{6}{0}$, $\binom{6}{1}$, ..., and $\binom{6}{6}$. Also, if we needed the binomial coefficients for $n = 21$, we could start with $\binom{21}{0} = 1$, and then pairwise adding the entries of the last row of Table II we would get $\binom{21}{1} = 1+20 = 21$, $\binom{21}{2} = 20+190 = 210$, $\binom{21}{3} = 190+1{,}140 = 1{,}330$, $\binom{21}{4} = 1{,}140+4{,}845 = 5{,}985$, and so on.

EXERCISES

1. In how many different ways can a motel chain select two of 11 sites for the construction of new motels? Use the first of the two formulas on page 23 and check your answer in Table II.

2. In how many ways can an accountant working for the Internal Revenue Service choose three of 10 tax returns for a special audit? Use the first of the two formulas on page 23 and check your answer in Table II. Also, if Mr. W. Scott's return is one of the 10, in what proportion of the possibilities will his return be audited?

3. Use the second of the two formulas on page 23 and Table I to find the number of ways in which a child can select six of the 12 rides in an amusement park. Also check the answer in Table II.

4. Use the second of the two formulas on page 23 and Table I to find the number of ways in which a student can choose four of eight elective subjects. Also check the answer in Table II.

5. In how many ways can a five-man committee be chosen from among the 18 teachers and the principal of an elementary school. Use the first of the two formulas on page 23 and check your answer in Table II. Also, determine in what percentage of these committees the principal will *not* be included.

6. Use the second of the two formulas on page 23 and Table I to find the number of ways in which a woman can pick four of the 15 dresses that a store carries in her size. Also check the answer in Table II.

7. In Exercise 19 on page 14 the reader was asked for the total number of ways in which a student can mark a true-false test which consists of 15 questions. In how many different ways can he or she mark this test and get
(a) three right and 12 wrong;
(b) six right and nine wrong;
(c) 12 right and three wrong?

8. Among Tom's friends there are eight boys and five girls. In how many ways can he invite two of the boys and three of the girls to a party? (*Hint*: multiply the respective number of possibilities in accordance with the rule on page 8.)

9. To fill a number of vacancies, the personnel manager of a company has to choose three secretaries from among six applicants and two bookkeepers from among four applicants. What is the total number of ways in which he can make his selection? (*Hint*: multiply the respective number of possibilities in accordance with the rule on page 8.)

10. On page 10 we showed that if a test consists of 12 multiple-choice questions, each permitting four alternatives, there are 16,777,216 ways in which a student can answer the test.

(a) In how many of these cases will the student get three correct answers and nine incorrect answers?

(b) In how many of these cases will the student get eight correct answers?

(c) Are there more possibilities for five correct answers or for six correct answers?

It is assumed here that each question has only one correct answer. (*Hint*: multiply the number of ways in which we can pick the questions that are answered correctly by the number of ways in which the other questions can be answered incorrectly.)

11. In hiring his staff, the manager of a wholesale food distributor has to choose five salesmen from among nine applicants, two buyers from among five applicants, and three secretaries from among 11 applicants. In how many different ways can he select his staff?

12. A shipment of 14 transistor radios contains one that is defective. In how many ways can we choose four of the radios for inspection so that

(a) the defective radio is not included;

(b) the defective radio is included?

Also find

(c) what percentage of all the choices contain the defective radio.

13. A carton of eggs contains two bad ones. In how many ways can we select three of the 12 eggs so that

(a) none of the bad ones are included;

(b) both of the bad ones are included;

(c) one of the bad ones is included?

Also find what percentage of all the selections contains

(d) at least one of the bad eggs;

(e) both of the bad eggs.

14. Among the 20 candidates for four positions on a city council, eight are Democrats, eight are Republicans, and four are Independents (that is, they have no official party affiliation).

(a) In how many ways can the four councilmen be chosen from among the 20 candidates?

(b) In how many ways can the four councilmen be chosen so that two are Democrats and two are Republicans?

(c) In how many ways can the four councilmen be chosen so that one is a Democrat, one is a Republican, and two are Independents?

15. Counting the number of outcomes in games of chance has been a popular pastime for many centuries. This was of interest not only because of the gambling that was involved, but also because the outcomes of games of chance were often interpreted as divine intent. Thus, it was just about a thousand years ago that a bishop in what is now Belgium determined that there are 56 different ways in which three dice can fall *provided one is interested only in the overall result and not in which die does what.* He assigned a virtue to each of these possibilities and each sinner had to concentrate for some time on the virtue which corresponded to his cast of the dice.

(a) Find the number of ways in which three dice can all come up with the same number of points.

(b) Find the number of ways in which two of the three dice can come up with the same number of points, while the third comes up with a different number of points.

(c) Find the number of ways in which all three of the dice can come up with a different number of points.

(d) Use the results of parts (a), (b), and (c) to verify the bishop's calculations that there are altogether 56 possibilities.

16. A judge of a traffic court has on his agenda three cases of speeding, one case of going through a red light, one case of reckless driving, and one case of driving while intoxicated. By type of violation only, in how many ways can this judge choose three of these cases to hear in a given morning. (*Hint*: determine separately the number of possibilities which include zero, one, two, and three cases of speeding.)

17. Rework part (d) of Exercise 10 on page 21, using the method illustrated on page 25.

18. Rework parts (d) and (e) of Exercise 11 on page 21, using the method illustrated on page 25.

19. Rework Exercise 12 on page 21, using the method illustrated on page 25.

20. Use Table II to determine (a) $\binom{16}{6}$, (b) $\binom{17}{12}$, (c) $\binom{20}{7}$, (d) $\binom{16}{13}$, and (e) $\binom{20}{15}$

21. Verify each of the following:

(a) $\binom{8}{5} = \binom{7}{4} + \binom{7}{5}$;

(b) $\binom{12}{9} = 4 \cdot \binom{11}{9}$;

(c) $\binom{12}{7} = \binom{11}{6} + \binom{11}{7}$;

(d) $7 \cdot \binom{10}{3} = 10 \cdot \binom{9}{3}$.

22. Expressing the binomial coefficients in terms of factorials and then performing whatever simplifications are necessary, prove that

(a) $\binom{n}{r} = \binom{n}{n-r}$;

(b) $\binom{n}{r} = \binom{n-1}{r-1} + \binom{n-1}{r}$;

(c) $\binom{n}{r} = \frac{n}{n-r} \cdot \binom{n-1}{r}$.

23. Rework Exercise 20 on page 14 by adding the number of ways in which we can take one of the coins, the number of ways in which we can take two of the coins, ..., and the number of ways in which we can take five of the coins.

24. Using the same argument as in Exercise 20 on page 14, it can be shown that there are 2^n ways in which we can select 0, 1, 2, ..., or all n of the elements of a set. Verify this result by showing that for special values of a and b the binomial theorem leads to the equation $2^n = \binom{n}{0} + \binom{n}{1} + \dots + \binom{n}{n}$. Also find the number of different ways in which we can select none, some, or all of the elements of a set with (a) 12 elements, and (b) 24 elements.

2

PROBABILITIES

INTRODUCTION

In everyday language we use words such as "probable," "likely," and other terms expressing uncertainties as a matter of course. "The train will probably be late." "More likely than not he will get a passing grade." "The chances are that the patient will recover." "It is very unlikely that his father will let him drive his new car." "In all probability the bank robber will be caught." "He is an odds-on choice to win the election." "There is but a slim chance that it will rain." All these statements express degrees of uncertainty, and they are all sufficiently vague so that we need not rack our brains to figure out precisely what they mean.

The situation is very much the same when we use words such as "quite," "fairly," "reasonably," or "about." "He is quite intelligent." "It is fairly hot outside." "His secretary is reasonably good at shorthand." "He is about as tall as his father." All these statements are vague too, but in contrast to those of the preceding paragraph, they can easily be made more precise. We can give an I.Q. test to the person who is supposed to be "quite" intelligent, we can take a look at a thermometer to see how hot it really is, we can check how many words per minute the secretary can take, and we can use a tape measure to check the height of father and son. To accomplish the same thing with the statements of the first paragraph, we would need a "yardstick" for probabilities, namely, some way of measuring degrees of uncertainty, and this is usually much easier said than done. To illustrate, let us consider the following situations:

CASE A. Suppose that a student has to take an English examination, in which he will be asked questions about one of twenty novels that were assigned as required reading. Of course, he does not know which one, and if he has read only two of the novels, we would say that the chances are poor that it will be one of the novels he read. Why do we say that the chances are poor? Well, there are 20 possibilities, and in only two of them he would have read the right novel. It stands to reason also that if the student has read five of the novels his chances would be better, and if he had read 10 of the novels we might say that there is a *fifty-fifty chance* that he will be asked questions about one of the novels he has read. All these judgments are based on a comparison of favorable and unfavorable possibilities, and if we wanted to go one step further, we could use the *proportion (or percentage) of favorable possibilities* as a measure of probability. Thus, if the student has read only two of the 20 novels, we would say that the probability is $\frac{2}{20} = 0.10$ (or 10 percent) that he will be asked questions about one of the novels he has read. Similarly, if he has read five of the novels, we would say that the probability is $\frac{5}{20} = 0.25$ (or 25 percent), and if he has read 10 of the novels, we would say that the probability is $\frac{10}{20} = 0.50$ (or 50 percent), and this is what we referred to earlier as a *fifty-fifty chance.*

CASE B. Suppose that Mr. Jones, who lives in a suburb, commutes to work by train, and each evening his wife meets him at the station in the family car. Mrs. Jones does not worry if she is a few minutes late, for as she tells her friends with whom she has been playing bridge, "The train is probably late, anyhow!" In this case, Mrs. Jones is not counting possibilities (either the train is late or it is not late, and that does not tell us anything), but in all likelihood she bases her judgment on *past experience*—she has met her husband at the station for years, and more often than not, the train has been late. To be more specific, she might say that the train will probably be late because it has been late in 68 of the 85 times she has gone to meet her husband during the last four months, which is 80 percent of the time. She might thus say that there is an *80 percent* chance that the train will be late. Similarly, a chief of police might say that there is a *64 percent chance* that a bank robber will get caught, basing this judgment on departmental statistics (which show that 64 percent of all bank robbers got caught); a doctor may tell us that there is an *excellent chance* that a patient will pull through, basing his opinion on the fact that in the past 96 percent of all patients having the same disease recovered; and the dean of a college may tell us that *the chances are not too good* that an entering freshman will graduate, basing this judgment on the fact that in the years 1965 through 1969 only 3,455 of 9,270 entering freshman, namely, $\frac{3{,}455}{9{,}270} \cdot 100 = 37$ percent, managed to graduate. *In all of these examples, the probability judgment was based on the percentage (or proportion) of the time an appropriate event has happened in the past, and in some instances this percentage (or proportion), itself, was used as a measure of probability.*

CASE C. Now let us consider the case of a businessman, who has just opened a bookstore near a college. What does he mean if he says that it is quite probable that he will make a profit during the first year? Of course, he could count possibilities as in Case A—either he will make a profit or he will not make a profit—but that would not make much sense. To some extent, he could perhaps rely on past experience as in Case B and base his judgment on the percentage of other bookstores near colleges that showed a profit during the first year, but if he is a good businessman he would also consider such things as business conditions in general, the nature of the competition, the availability of financing in case it is needed, the quality of management as well as labor, the cost of rent and insurance, and quite a few other factors which have proven to be of relevance. Thus, the opinion that he will "probably" make a profit during the first year will have to be based on a more or less subjective evaluation of all these factors, and this makes it difficult to construct a "yardstick" for this kind of probability. One way of measuring such subjective evaluations of a person's chances (in this case the businessman's chances of making a profit during the first year) is to put the whole thing on a "put-up or shut-up" basis. For instance, we may ask the businessman whether he would be willing to bet $300 against $200 that he will make a profit during the first year. If the the answer is "Yes," would he also be willing to bet $500 against $200 or even $1,000 against $100? Or we might ask him what he considers "fair odds," where "odds" means the ratio of the amount he is willing to risk to the amount he stands to win. Clearly, if he is pretty confident that he will make a profit during the first year, he may well be willing to bet $500 against $200, or if he is even more confident, he may be willing to bet $800 against $200. On the other hand, if he feels that there is less than a fifty-fifty chance, he may be willing to bet only $100 against $200, which means that he will win $200 if the bookstore makes a profit during the first year, but lose only $100 in case of the more likely event that the bookstore will not make a profit during the first year. Odds like these, which reflect what a person considers a fair risk, are sometimes referred to as **relative probabilities.** Thus, if the businessman feels that betting $800 against $200 is fair, the word "probable" in his original statement on page 000 could be interpreted as a relative probability of 8 to 2 (or 4 to 1, which is the same); on the other hand, if he felt that in all fairness he can bet only $100 against $200, this could be interpreted as a relative probability of 1 to 2. (Later in this chapter we shall see how such relative probabilities can be converted into "ordinary" probabilities, namely, probabilities which measure degrees of uncertainty in terms of a single number.)

The purpose of this discussion has been to demonstrate that, depending on the circumstances, different "yardsticks" may be needed to measure uncertainties. What makes these circumstances different is the *nature* of the information on which the probability judgments are based. In Case A we had no idea how the student's instructor made up the test or how the student chose the two novels which he read—we knew only that there were 20 possibilities, in two of which the student would be asked questions about a novel he has read. In Case B, on the other hand, the probability judgment was based entirely on observations of what happened in the past, and in Case C it was based on a subjective combination of various kinds of direct and collateral information. As the reader can well imagine, there is much more to these probability yardsticks than we have discussed—indeed, we shall see later in this chapter that their applications are limited that they are controversial, and that they raise all sorts of philosophical questions.

THE CLASSICAL CONCEPT

When we discussed Case A on page 33, it may have occurred to the reader that we failed to mention that the 20 possibilities had to be regarded as *equally likely,* and that without this assumption the whole argument would fall apart. This is correct, of course, and it points to one of the main shortcomings of the so-called **classical concept of probability**. This concept is the oldest historically, and it was developed originally in connection with games of chance, which lent themselves most readily to the big step *from possibilities to probabilities.* Take a die, for example. It is a cube with six faces, and by virtue of its symmetry it was argued (and still is) that one face should be just as likely to come up as any other. Similarly, if we flip a coin there are two possibilities, *heads* or *tails,* which we regard as equally likely, and if we use the spinner of Figure 2.1 there are presumably 10 equally likely possibilities. Then there is the usual deck of 52

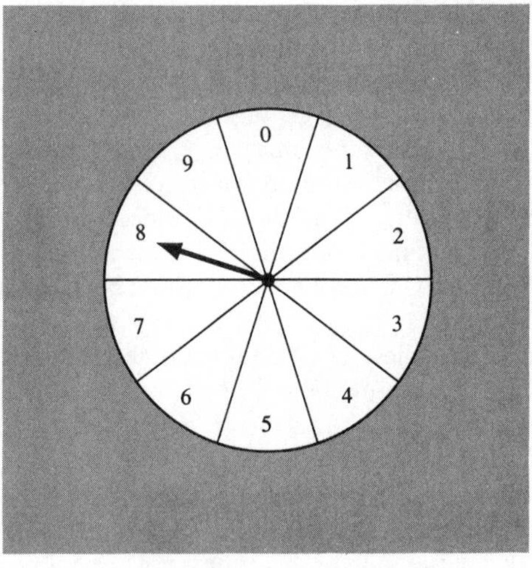

FIGURE 2.1 Spinner.

playing cards, where we assume that one card is as likely to be drawn as another (provided the deck has been shuffled to one's satisfaction). What all these devices have in common is that their symmetry, or structure, makes it reasonable to assume that all possible outcomes are equally likely. This kind of argument is typical of classical, prescientific thought, where properties of objects were deduced from their structure and their shape instead of their "performance."

To provide a probability "yardstick" which can be used whenever all possibilities are equally likely (or, at least, when we can assume that they are equally likely), we proceed as in Case A on page 33 and use the following rule:

If there are n equally likely possibilities, of which one must occur and s are regarded as favorable, namely, as constituting a success, then the probability of a success is given by the ratio $\frac{s}{n}$.

In the application of this rule it should be understood that the terms "favorable" and "success" are used rather loosely. What is favorable to one player is unfavorable to his opponent, and what is a success from one point of view is a failure from another. Thus, the terms "favorable" and "success" can be applied to any particular kind of outcome with which we happen to be concerned, even if it turns out that "favorable" means that a television set does not work, or a "success" is a student's failing an exam. As the reader may have guessed, this usage of the terms "favorable" and "success" dates back to the days when probability theory was used only in connection with games of chance.

Following the rule of the preceeding paragraph, we find, for example, that the probability of rolling a 1, 2, or 3 with a die is $\frac{3}{6} = \frac{1}{2}$, the probability of drawing an ace from a (well-shuffled) deck of 52 playing cards is $\frac{4}{52} = \frac{1}{13}$, and the probability of getting a 2, 4, 6, or 8 with the spinner of Figure 2.1 is $\frac{4}{10} = \frac{2}{5}$. We also find that the probability for each face of a die is $\frac{1}{6}$, the probabilities for *heads* and *tails* are both $\frac{1}{2}$, the probability for each of the 52 playing cards is $\frac{1}{52}$, and in general that

If there are n equally likely possibilities, of which one must occur, then each possibility has the probability $\frac{1}{n}$.

An immediate consequence of this way of measuring uncertainties is that *when all the possibilities are regarded as favorable, the certainty of a favorable outcome has the probability* $\frac{n}{n} = 1$, and *when none of the possibilities is regarded as favorable, the possibility (that is, impossibility) of a favorable outcome has the probability* $\frac{0}{n} = 0$. Otherwise, all probabilities are fractions between 0 and 1, with larger values being indicative of a greater proportion of favorable possibilities and, hence, a better chance of success. More specifically, the probability of a success is always *proportional* to the number of favorable possibilities, and this certainly makes sense—when there are twice as many favorable possibilities, namely, twice as many "chances" of success, the probability of success should also be twice as large.

Another immediate consequence of this way of measuring uncertainties is that *the sum of the probabilities of success and failure must always equal 1.* To prove this, we have only to observe that when the number of favorable cases is s, the number of unfavorable cases is $n-s$, the probability of success is $\frac{s}{n}$, the probability of failure is $\frac{n-s}{n}$, and $\frac{s}{n}+\frac{n-s}{n} = \frac{s+n-s}{n} = \frac{n}{n} = 1$. For instance, eight of the 52 cards in an ordinary deck of 52 playing cards are kings or queens, the other 44 cards are neither kings nor queens, the probability of drawing a king or a queen is $\frac{8}{52}$, the probability of *not* drawing a king or a queen is $\frac{44}{52}$, and $\frac{8}{52}+\frac{44}{52} = 1$.

Since the classical concept of probability requires that we count possibilities (the total number of possibilities as well as the number of favorable possibilities), it provides many applications of the methods of Chapter 1. To give a few examples, let us first determine the probability that two cards drawn from an ordinary deck of 52 playing cards will both be red. According to what we learned in Chapter 1, the total number of possibilities is $\binom{52}{2}$, the number of favorable possibilities is $\binom{26}{2}$ since half of the 52 cards are red and the other half are black, and it follows that the probability of drawing two red cards is

$$\frac{\binom{26}{2}}{\binom{52}{2}} = \frac{25}{102}$$

As a second example, let us find the probability of getting exactly two 3's in three rolls of a die. In this case the total number of possibilities is $6 \cdot 6 \cdot 6 = 216$, since there are six possibilities for each roll of the die, and the number of favorable possibilities is the *product* of $\binom{3}{2}$ and 5, which are, respectively, the number of ways in which we can choose the two rolls on which the die comes up 3, and the number of possibilities for the other roll of the die. Thus, the probability is

$$\frac{\binom{3}{2}\cdot 5}{216} = \frac{3\cdot 5}{216} = \frac{5}{72}$$

Although equally likely possibilities are found mostly in games of chance, the classical probability concept applies also in a great variety of situations where gambling devices are used to make so-called **random selections**—say, when new offices are assigned to an insurance company's agents by lot, when subjects for debate are assigned to students by drawing slips of paper out of a hat, or when an issue is decided by the flip of a coin.

Suppose, for instance, that a television manufacturer requires that four of the 20 sets in each production lot must be inspected before they are shipped. If the sets are all satisfactory the whole lot is shipped, but if they are not all satisfactory the remaining 16 sets are also inspected. It is assumed that the four sets are chosen at random. What we would like to know is the probability that a lot will "pass" this inspection when actually one of the 20 sets is defective. Clearly, the total number of possibilities is $\binom{20}{4}$, the total number of favorable possibilities is $\binom{19}{4}$, and the answer is

$$\frac{\binom{19}{4}}{\binom{20}{4}} = \frac{3{,}876}{4{,}845} = 0.80$$

To consider another example, let us refer back to the illustration on page 22, where we showed that a housewife can choose three of the 12 cheeses stocked by a supermarket in 220 different ways. Now, if we want to find the probability that she will choose neither Cheddar nor Gruyère, two of the 12 cheeses stocked by the store, we can argue that the number of favorable possibilities is $\binom{10}{3}$, and, hence, that the desired probability is

$$\frac{\binom{10}{3}}{220} = \frac{120}{220} = \frac{6}{11}$$

In Exercise 14 on page 44 the reader will be asked to continue this example by showing that the probability of the housewife not choosing Gruyère is $\frac{3}{4}$, and the probability of her choosing Cheddar as well as Gruyère is $\frac{1}{22}$. Of course, all these calculations require the assumption that all of the possibilities are equally likely, which is debatable in this case.

Finally, let us determine the probability that a student's two business texts (one on management and one on advertising) will be next to each other when he haphazardly puts his eight textbooks on a shelf. In this case we are dealing with permutations, and we find that the total number of possibilities (the total number of ways in which the books can be arranged on the shelf) is 8!. So far as the favorable possibilities are concerned, their number is $7 \cdot 2 \cdot 6!$, since there are seven ways in which we can choose the two adjacent places for the two business books (first and second from the left, second and third from the left, and so on), there are two ways in which the two business books can be put into these two places, and there are 6! ways in which we can arrange the other six books. Thus, the desired probability is

$$\frac{7 \cdot 2 \cdot 6!}{8!} = \frac{1}{4}$$

Although the classical probability concept has great intuitive appeal, it also has some very serious shortcomings. First of all, it can be used only when we are dealing with equally likely possibilities, and it cannot be used, therefore, when we speak of the probability that it will rain, the probability that we will have a flat tire while driving from Los Angeles to San Francisco, or the probability that an airplane will arrive on time. Surely, it would be nonsensical to say that since it will be either sunny or overcast, or there will be rain, hail, or snow, the probability of rain is $\frac{1}{5}$. Similarly, it would be silly to say that the probability is $\frac{1}{3}$ that a plane will arrive on time, because there are only the three possibilities: early, on time, or late.

As we have already pointed out, the classical concept applies mainly to games of chance, or when gambling devices are used to select a sample, distribute objects, or make decisions. Even then, however, we can never be *really sure* that the possibilities are all equally likely—a die can be loaded, a coin can be off balance, and some of the cards in a deck may stick together. To forestall this kind of criticism, we often hedge our probability statements by inserting the word "if." For instance, in the first example on page 38 we might rephrase the result by saying that "if each card has the same chance of being selected, the probability of drawing two red cards is $\frac{25}{102}$." Similarly, in the second example we might rephrase the result by saying that "if the die is perfectly balanced, the probability of getting exactly two 3's is $\frac{5}{72}$." This really does not solve anything, for we are simply leaving it open whether the respective possibilities can all be regarded as equally likely.

Another criticism which is often raised against the classical concept is that the rule on page 37, according to which probabilities are given by the ratio $\frac{s}{n}$, does not really provide a *definition* of probability. The reasoning is *circular*, it is argued, because we are "defining" probability with reference to equally likely, or equiprobable, possibilities, and we cannot very well know what is meant by "equiprobable" unless we already know what is meant by "probability." To answer this particular criticism of the classical concept, we could point to other branches of mathematics, where we always leave some terms *undefined*. For instance, we speak of numbers in arithmetic without ever giving a *rigorous* definition, say, of the number *two*, and we do the same thing with points in elementary geometry and sets in modern mathematics. All these terms, or concepts, are supposed to be *understood without explanation*, and in connection with the classical concept of probability we can, similarly, treat "equally likely" as an undefined term. If the reader feels that this is a bit contrived, let us point out that it is, indeed, controversial, and that *we could manage without the classical concept by using one of the other concepts which we shall discuss later in this chapter*. Historically, though, the classical probability concept is of importance, and it is difficult to deny its appeal in connection with games of chance.

PROBABILITIES AND ODDS

In the discussion of Case C on page 34, we suggested that the odds at which a person is willing to bet on something are indicative of the likelihood that it will occur. Of course, in that example the likelihood, or probability, was a personal matter, as it told us how a particular person felt about the uncertainties he had to face. In connection with the classical concept of probability, the odds that something will occur, namely, the *odds for success*, have nothing to do with a person's willingness to bet—they are defined simply as *the ratio of the number of favorable possibilities to the number of unfavorable possibilities*. Similarly, the ratio of the number of unfavorable possibilities to the number of favorable possibilities gives the *odds for failure*, or the *odds against success*. In connection with the examples on page 37, we can thus say that the odds of rolling a 1, 2, or 3 with a die are 3 to 3, the odds of drawing an ace from a deck of playing cards are 4 to 48, and the odds of getting a 2, 4, 6, or 8 with the spinner of Figure 2.1 are 4 to 6. Correspondingly, the odds *against* rolling a 1, 2, or 3 with a die are also 3 to 3, the odds *against* drawing an ace are 48 to 4, and the odds *against* getting a 2, 4, 6, or 8 with the spinner are 6 to 4.

It is customary to quote odds as ratios of positive integers having no common factors, and to quote them in such a way (namely, *for or against*) that *the larger number comes first.* Thus, in the second of the above examples we would say that the odds *against* drawing an ace are 12 to 1; similarly, in the example on page 39 we would say that the odds are 6 to 5 that the housewife will choose neither Cheddar nor Gruyere.

As should be apparent, the classical concepts of probability and odds are closely related. If the number of favorable and unfavorable possibilities are, respectively, s and $n-s$, the odds for success are s to $n-s$, and this is the same as the ratio of $\frac{s}{n}$ to $\frac{n-s}{n}$, which can also be written as p to $1-p$, where $p = \frac{s}{n}$ is the probability of success and $1-p = 1-\frac{s}{n} = \frac{n-s}{n}$ is the probability of failure. Indeed, this is a good way of interpreting odds—*the odds for success are the ratio of the probability of success to the probability of failure.* Thus, if the probability of success is $\frac{8}{13}$, the probability of failure is $1-\frac{8}{13} = \frac{5}{13}$, and the odds for success are $\frac{8}{13}$ to $\frac{5}{13}$, or simply 8 to 5.

With this method we can always convert probabilities to odds, except for $p = 0$ or $p = 1$ when the odds are *undefined.* To proceed the other way, from odds to probabilities, we can use the following argument: if the odds for success are a to b, where a and b are positive integers, and p is the probability of success, then a and b are in the same ratio as p and $1-p$, or in other words $\frac{a}{b} = \frac{p}{1-p}$. As the reader will be asked to verify in Exercise 30 on page 47, the solution of this equation for p yields

$$p = \frac{a}{a+b}$$

For instance, if the odds for success are 13 to 7, the probability of success is $\frac{13}{13+7} = \frac{13}{20}$, and if the odds against success are 8 to 3, the probability of failure is $\frac{8}{8+3} = \frac{8}{11}$ and the probability of success is $\frac{3}{3+8} = \frac{3}{11}$. Thus, we have learned how to convert probabilities to odds and odds to probabilities, and as we shall see later, the relationships we have established here between probabilities and odds will apply also when uncertainties are measured as in Cases B and C, which we discussed in the beginning of this chapter.

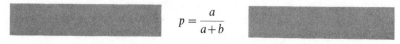

EXERCISES

1. When we roll a balanced die, what are the probabilities of getting (a) 1 or 3; (b) an even number; (c) a multiple of 3?

2. When one card is drawn from a well-shuffled deck of 52 playing cards, what are the probabilities of getting (a) a black king; (b) a jack, queen, king, or ace; (c) a card which is neither a jack nor a queen; (d) a heart or a spade; (e) a 4, 5, 6, 7, or 8 of any suit?

3. If H stands for *heads* and T for *tails,* the four possible outcomes for two flips of a coin are HH, HT, TH, and TT. Assuming that these four possibilities are equally likely, what are the respective probabilities of getting 0, 1, or 2 heads?

4. If H stands for *heads* and T for *tails,* the eight possible outcomes for three flips of a coin are HHH, HHT, HTH, THH, HTT, THT, TTH, and TTT. Assuming that these eight possibilities are equally likely, what are the respective probabilities of getting zero, one, two, or three heads?

5. A bowl contains 12 red beads, 10 white beads, 25 blue beads, and three black beads. If one bead is drawn at random, what is the probability that it will be (a) blue; (b) red or white; (c) black; (d) neither white nor black?

6. A hat contains 100 slips of paper numbered from one to 100. If one of these slips of paper is drawn at random, find the probabilities of getting a number which is (a) even; (b) greater than 60; (c) divisible by 11.

7. If one letter is chosen at random from the word "esteem," what is the probability that it will be an "e"?

8. If two letters are chosen at random from the word "housewarming," what is the probability that (a) they will both be vowels; (b) one will be a vowel and one will be a consonant?

9. If Mr. and Mrs. Jones and two friends sit down to play bridge and they draw lots for their chairs, what is the probability that Mr. Jones will *not* sit opposite Mrs. Jones?

10. When two cards are drawn from an ordinary deck of 52 playing cards, as in the example on page 38, what are the probabilities of getting (a) two aces; (b) two spades, of which there are 13 in the deck; (c) a king and a queen?

11. When four cards are drawn from an ordinary deck of 52 playing cards, show that the probability of getting one card of each suit is $\frac{2,197}{20,825}$.

12. Proceeding as in the illustration on page 38 find the probabilities of getting zero 3's, one 3, and three 3's in three rolls of a balanced die. Also verify that the sum of the three results obtained here and the one obtained on page 39 is 1.

13. With reference to the example on page 39, what are the probabilities that a lot will pass the inspection when
(a) two of the 20 sets are defective;
(b) three of the 20 sets are defective?

14. With reference to the illustration on page 40, check the values given for the probability that the housewife will not choose Gruyère cheese and the probability that she will choose Cheddar as well as Gruyère.

15. With reference to the example on page 40, what is the probability that there will be exactly one other book between the two business books?

16. A hoard of medieval silver coins discovered in Belgium included 120 struck for Charles the Bold and 60 struck for Philip the Good.
(a) If a person is allowed to pick one of these coins at random, what is the probability that he will get a coin struck for Charles the Bold?
(b) If a person is allowed to pick two of these coins at random what are the probabilities that he will get (i) two coins struck for Charles the Bold, (ii) two coins struck for Philip the Good; and (iii) one coin struck for each of these rulers?

17. Assuming that the awards are distributed at random among the 14 boys and 11 girls in the example on page 9, find the probabilities that
(a) both awards will be given to boys;
(b) both awards will be given to girls;
(c) one award will be given to a boy and one award will be given to a girl.
Is it a coincidence that the sum of the three probabilities is 1?

18. A student takes a multiple-choice test consisting of 10 questions, each permitting a choice of three alternatives. If he answers each question by rolling a die, checking the first alternative when he gets a 1 or 2, the second alternative when he gets a 3 or 4, and the third alternative when he gets a 5 or 6, what is the probability that he will
(a) answer each question correctly;
(b) answer each question incorrectly;
(c) answer three questions correctly and seven questions incorrectly;

(d) answer five questions correctly and five questions incorrectly?

19. With reference to Exercise 19 on page 14, what are the probabilities that a student who answers each of the 15 questions by flipping a coin (*heads* means "true" and *tails* means "false") will
(a) answer all the questions correctly;
(b) answer exactly five of the questions correctly;
(c) answer exactly seven of the questions correctly;
(d) answer none of the questions correctly?

20. Assuming in Exercise 1 on page 20 that Buffalo, N. Y., is one of the eight cities and that the route of three cities is chosen at random for each trip, find the probability that the salesman will not visit Buffalo on his next trip.

21. With reference to Exercise 4 on page 20, suppose that the student randomly takes two of the books home with him during the Christmas vacation. What are the probabilities of his taking
(a) two business books;
(b) two foreign language books;
(c) one of each kind?
Is it a coincidence that the sum of the three probabilities is 1?

22. With reference to the example on page 23, what are the probabilities that the survey of eight of the 12 largest cities in the United States will
(a) include New York City;
(b) include Los Angeles and Chicago;
(c) not include Houston?
On what assumption are these probabilities based?

23. If in Exercise 9 on page 29 the personnel manager makes his selection by drawing lots, what is the probability that his choice will include Miss Brown, Miss Adams, and Mr. Smith, where the first two are applicants for secretarial jobs and the third is an applicant for the bookkeeping jobs?

24. With reference to Exercise 13 on page 30, if the three eggs are selected at random, what are the odds that (a) none of the bad ones will be included; (b) both of the bad ones will be included; (c) only one of the bad ones will be included?

25. With reference to Exercise 14 on page 30, suppose that the four positions on the city council are filled by lot.
(a) What are the odds that two of the chosen candidates will belong to each of the two major parties?
(b) What are the odds against the possibility that all the positions are filled by Independents?

(c) What is the probability that at least one of the Democrats will be chosen?

[*Hint*: in part (c) find the probability of success by subtracting from 1 the probability of failure.]

26. Convert each of the following probabilities to odds:

(a) The probability of getting two heads in six flips of a balanced coin is $\frac{15}{64}$.

(b) The probability of drawing a one-eyed jack from an ordinary deck of 52 playing cards is $\frac{2}{52}$.

(c) The probability that the last digit of a car's license plate is a 3, 4, 5, 6, 7, 8, or 9 is $\frac{7}{10}$.

27. Convert each of the following probabilities to odds:

(a) A secretary is supposed to send three of six letters by special delivery. If she arbitrarily puts the special delivery stamps on three of the letters, the probability that she puts all these stamps on the right letters is $\frac{1}{20}$.

(b) If a teacher arbitrarily selects two of eight students to recite parts of a poem, the probability that Bill (who is one of these eight students) will be chosen is $\frac{1}{4}$.

(c) A person has eight \$1 bills, five \$5 bills, and one \$10 bill in his wallet. If he randomly pulls out three of the bills, the probability that they are not all \$1 bills is $\frac{11}{13}$.

28. Convert each of the following odds to probabilities:

(a) The odds against rolling "7 or 11" with a pair of dice are 7 to 2.

(b) If an urn contains 22 black balls and seven white balls, the odds for drawing two black balls in succession are 33 to 25.

(c) When we flip a coin four times, the odds against getting one *heads* and three *tails* are 3 to 1.

29. Convert each of the following odds to probabilities:

(a) On a tray there are six pieces of apple pie and four pieces of cherry pie. If a waitress arbitrarily takes two of these pieces of pie and gives them to customers who ordered cherry pie, the odds are 13 to 2 that she is making a mistake.

(b) Four businessmen having lunch together hang up their coats before they sit down to eat. If afterwards each of them arbitrarily grabs one of the coats, the odds are 3 to 1 against two getting the right coat and two getting a wrong coat, and the odds are 5 to 3 that at least one of them gets the right coat.

(c) If we arbitrarily arrange the letters in the word "nest," the odds are 5 to 1 that we will not get a meaningful word in the English language.

30. Verify that if the equation $\frac{a}{b} = \frac{p}{1-p}$ is solved for p, the result is $p = \frac{a}{a+b}$.

31. Explain why we cannot speak of the odds for failure or success when the probability of success is 0 or 1.

32. With reference to the discussion of Case A on page 33, explain under what circumstances it would be quite unreasonable to maintain that all the possibilities are equally likely.

33. With reference to the illustration on page 22, where the housewife chooses three of the 12 cheeses, explain why under normal conditions it would be quite unreasonable to assume that all the possibilities are equally likely.

34. Some philosophers have argued that if we have absolutely no information about the likelihood of the different possibilities, it is reasonable to regard them all as equally likely. This is sometimes referred to as the "principle of *equal ignorance*." Discuss the argument that human life either does or does not exist elsewhere in the universe, and since we really have no information one way or the other, the probability that human life exists elsewhere in the universe is $\frac{1}{2}$.

THE FREQUENCY INTERPRETATION

Among the various concepts of probability, most widely held is the **frequency interpretation**, according to which *the probability of an event (happening, or outcome) is interpreted as the proportion of the time that similar events will occur in the long run.* Thus, if we say that there is a probability of 0.86 that a jet from Chicago to Los Angeles will arrive on time, this means that such flights actually arrive on time about 86 percent of the time. Also, if the weather bureau predicts that there is a 30 percent chance for rain, namely, a probability of 0.30, this means that under the same weather conditions it will rain about 30 percent of the time. More generally, we say that an event has a probability of, say, 0.90, in the same sense in which we might say that in cold weather our car will start on the first try about 90 percent of the time. *We cannot guarantee what will happen on any particular occasion—the car may start on the first try and then it may not—but if we kept records for a long time, we should find that the proportion of "successes" is very close to 0.90.*

An immediate consequence of the frequency interpretation is that we can never actually *know* the value of a probability; after all, who lives long enough to know precisely what happens in the long run? Although this is true, it is not as serious as it may seem. Granted that we can never really know the *exact* value of a probability, we can nevertheless *estimate* it by observing how often (what part of the time) similar events have occurred in the past. For instance, if airline records show that over a certain period of time 688 of 800 jets from Chicago to Los Angeles arrived on time, we estimate the probability that any one flight from Chicago to Los Angeles (say, the next one) will arrive on time as $\frac{688}{800} = 0.86$. Similarly, if 1,512 of 2,439 washing machines produced by a certain firm required repairs within the first year, we estimate the probability that any one of their machines will require repairs within the first year as $\frac{1{,}512}{2{,}439} = 0.62$. Also, in connection with Case B on page 34, we could have estimated the probability that a freshman entering the given college will graduate as $\frac{3{,}435}{9{,}270} = 0.37$.

When probabilities are thus estimated, it is only reasonable to ask whether the estimates are any good. The answer, which is "Yes," is supported by a remarkable law called the **Law of Large Numbers**, which will be proved and discussed in some detail in Chapter 8. Informally, this law can be stated as follows:

> **If the number of times a situation is repeated becomes larger and larger, the proportion of successes will tend to come closer and closer to the actual probability of success.**

To illustrate how this works, let us refer to the familiar example of repeatedly flipping a coin. If we observe the accumulated proportion of successes (say, *heads*) after every fifth flip and plot it graphically as in Figure 2.2, it can be seen that although it fluctuates, the proportion comes closer and closer to $\frac{1}{2}$, the probability of heads for each flip of the coin. The best way to develop an understanding of the Law of Large Numbers is through experimentation (that is, repeatedly flipping a coin, rolling dice, and so forth), and this is why the time needed for Exercises 26, 27, and 28 on page 57 will be well spent.

In accordance with the frequency interpretation, probabilities are again numbers on the interval from 0 to 1, with 1 representing the *certainty* of something which happens 100 percent of the time, and 0 indicating the *impossibility* of something which cannot happen, namely, something which happens 0 percent of the time. Another property which carries over from the classical concept is that *the sum of the probabilities of success and failure must always equal 1.* For instance, if we say that the probability for

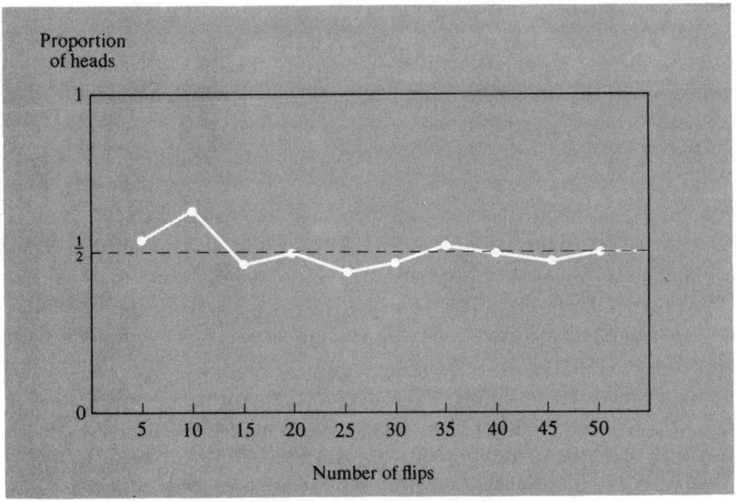

FIGURE 2.2 Coin-tossing experiment.

rain on a given day is 0.30, we mean that it will rain on days like that 30 percent of the time. Therefore, it will not rain on days like that 70 percent of the time, the probability that it will not rain is 0.70, and $0.30+0.70 = 1$. Similarly, if the probability is 0.64 that any applicant for a driver's license will pass the road test, this means that 64 percent of all applicants will pass; thus, 36 percent will not pass, the probability that an applicant will not pass is 0.36, and $0.64+0.36 = 1$. All we are really saying is that the two proportions *pro and con* add up to 1.

On page 42 we said that the odds for a success are best looked upon as the ratio of the probability of success to the probability of failure, and this is precisely how we shall *define* odds in connection with the frequency interpretation.* Thus, if the probability is 0.75 that on our way to work we have to wait more than 10 minutes for a bus, then the odds are 0.75 to 0.25, or 3 to 1, that on any given occasion we will have to wait more than 10 minutes for a bus on our way to work. Also, if the probability for rain is 0.30, the odds that it will rain are 0.30 to 0.70, or 3 to 7, and hence 7 to 3 that it will *not* rain. When we said on

* As in the classical probability concept, this definition will not apply when the probability of success is either 0 or 1.

page 47 that the probability is 0.90 that in cold weather our car will start on the first try, we can now interpret this statement as follows: we cannot guarantee what will happen at any particular occasion, but we would consider it *fair* to give odds of 9 to 1 (say, bet \$9 against \$1) that the car will start on any given occasion on the first try. This would be *fair,* or *equitable,* for we would win \$1 about 90 percent of the time, lose \$9 about 10 percent of the time, and hence, *can expect to break even in the long run.* Note also that if we want to convert odds into probabilities, we proceed as on page 42; that is, we use the formula $p = \frac{a}{a+b}$, where the odds for success are a to b and the probability of success is p. The argument is the same, regardless of whether we look upon probabilities in the classical sense, or as "proportions in the long run."

Having defined probabilities in terms of what happens to similar events in the long run, let us examine briefly whether it is at all meaningful to talk about the probability of something which occurs just once. For instance, can we ask for the probability that Mrs. H. E. Brown's broken wrist will heal within a month, or the probability that a certain major-party candidate will win an upcoming presidential election? If we put ourselves in the position of Mrs. Brown's doctor, we could check medical records, discover, say, that in the past such fractures have healed within a month 31 per cent of the time, and apply this figure to Mrs. Brown. This may not be of much comfort to Mrs. Brown, but it does provide a meaningful probability statement concerning her broken wrist—the probability that it will heal within a month is 0.31. Thus, *when something happens just once, the frequency interpretation leaves us no choice but to refer to a set of similar situations.*

It is easy to see how this can lead to complications, for the choice of what is "similar" is often neither obvious nor straightforward. With reference to Mrs. Brown's wrist, we might consider as "similar" only those cases where the fracture was in the same (left or right) wrist, we might consider only those cases in which the patients were exactly her age, or we might consider only those cases in which the patients were also of the same height and weight. Ultimately, this is a matter of individual judgment, and it is by no means contradictory that we can thus arrive at different probability estimates concerning one and the same event. It should be observed, however, that the more we narrow things down, the less information we will have for estimating the corresponding probability; that is, we may have more information about the *particular case,* but at the same time less information about *similar cases.*

So far as the second example is concerned, the one concerning the presidential election, suppose we ask a pollster "how sure" he is that the given candidate will actually win the election. If he replies that he is, say,

"95 percent sure" (that is, he assigns the candidate a success probability of 0.95), this surely can't mean that the candidate should win about 95 percent of the time, if he ran for office a great many times. *No, it means that the pollster based his conclusions (that the given candidate will win) on methods which in the long run will work 95 percent of the time.* In this sense, many of the probabilities which we use to express our faith in predictions, judgments, or decisions, are simply *"success ratios" that apply to the method we have employed.*

SUBJECTIVE PROBABILITIES

A point of view which is currently gaining in popularity is to interpret probabilities *subjectively,* as in Case C on page 34. This method of dealing with uncertainties works nicely (and is certainly justifiable) in situations where there is very little *direct evidence,* and there may be no choice but to consider a combination of collateral (indirect) information, "educated guesses," and perhaps intuition and other subjective factors. As we pointed out on page 35, a businessman who has just opened a new bookstore may well base his chances for success on the experience of other bookstores, business conditions in general, the availability of labor, and numerous other factors, which all in some way contribute to the overall picture. We also suggested that his feeling about the uncertainty of success (namely, a profit during the first year) can be measured by the odds at which he would be willing to bet. If he is pretty confident that he will make a profit during the first year, he may be willing to bet $500 against $200 that this will be the case; if he is even more confident, he may be willing to bet $800 against $200; and if he feels that his chances are not so good, he may be willing to bet only $150 against $200. Now we can go one step further and convert these odds into probabilities with the same formula which we used on pages 42 and 50 in connection with the classical probability concept and the frequency interpretation. This leads to the following definition:

If a person feels that a to b are fair odds for betting on a success of a given kind, he is, in fact, assigning to such a success the subjective, or personal probability

$$p = \frac{a}{a+b}$$

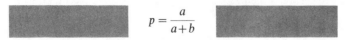

Thus, if the businessman of Case C felt that betting $500 against $200 (on his making a profit during the first year) is a fair bet, he would be assigning this success a personal, or subjective probability of $p = \frac{500}{500+200} = \frac{5}{7}$.

One disadvantage of this method of measuring subjective probabilities is that it is sometimes (indeed, more often than not) difficult to pin down exactly what a person considers a fair bet. The best approach is to put things on a "put-up or shut-up" basis. Suppose, for instance, we ask a student whether he would be willing to give us 7 to 1 odds (namely, bet his \$7 against our \$1) that he will get an A in a final examination. Suppose, furthermore, that he turns us down and that we then ask him whether he is willing to give us 4 to 1 odds (namely, bet his \$4 against our \$1). If he accepts this second bet, we can then argue as follows: If he had considered the 7 to 1 odds fair, he would have assigned his getting an A in the final examination a probability of $\frac{7}{7+1}$, but since he turned them down, this means that he feels that his chances are *not that good,* namely, that $p < \frac{7}{8}$, where p is the probability that he will get an A in the exam. If he had considered the 4 to 1 odds fair, he would have assigned his getting an A in the final examination a probability of $\frac{4}{4+1} = \frac{4}{5}$, and since he accepted these odds, this means that he feels that his chances are *at least that good,* namely, that $p \geqq \frac{4}{5}$.* Even though we have not pinpointed the exact value of the probability, we have shown that it lies on the interval from $\frac{4}{5}$ to $\frac{7}{8}$, including $\frac{4}{5}$ but not $\frac{7}{8}$. Symbolically, we have arrived at the result that $\frac{4}{5} \leqq p < \frac{7}{8}$, and if we wanted to narrow this down, we could ask the student whether he would be willing to give us odds of, say, 5 to 1 or 6 to 1 (see Exercise 15 on page 56).

When subjective probabilities are measured in this way, they are again numbers on the interval from 0 to 1. Also, the surer a person is that something will occur, the better are the odds he would be willing to give—say, 100 to 1, 1,000 to 1, or perhaps even 1,000,000 to 1. The corresponding probabilities are $\frac{100}{100+1}$, $\frac{1,000}{1,000+1}$, and $\frac{1,000,000}{1,000,000+1}$ (or approximately 0.99, 0.999, and 0.999999), and it can be seen that *the more certain a person is that something will occur, the closer to 1 will be the probability which he assigns to it.* Actually, if a person is *certain* of the outcome, he should be willing to give any odds (after all, he cannot possibly lose), and we again assign to certainty the probability 1. Also, if a person feels that an outcome

* In case the reader is not familiar with inequality signs, $p < \frac{7}{8}$ means that p is less than $\frac{7}{8}$, and $p \geqq \frac{4}{5}$ means that p is greater than or equal to $\frac{4}{5}$.

is so unlikely that he would be unwilling to bet on it even if he were given odds of 1,000 to 1, or even 1,000,000 to 1, he would be assigning this outcome a probability less than $\frac{1}{1{,}001}$ (approximately 0.001), or less than $\frac{1}{1{,}000{,}001}$ (approximately 0.000001). Thus, *the more certain a person is that something will not occur, the closer to 0 will be the probability which he assigns to it.* Finally, if a person is *certain* that something cannot occur, he would not bet on it at any odds, and we again assign to impossibilities the probability 0.

The fact that the probabilities of success and failure add up to 1 applies also to subjective probabilities. If it is fair to bet a dollars against b dollars that something will occur, it would also be fair to bet b dollars against a dollars that it will not occur. Thus, the probabilities of success and failure are, respectively, $\frac{b}{a+b}$ and $\frac{b}{b+a}$, and $\frac{a}{a+b}+\frac{b}{b+a}=\frac{a+b}{a+b}=1$. In all three cases, therefore, the probability of failure can be obtained by subtracting from 1 the probability of success.

As it gets confusing when we use the same term in different connotations, let us point out that we have actually given three different definitions of what we mean by "odds." In the classical concept, odds are the ratio of the number of favorable cases to the number of unfavorable cases; in the frequency interpretation, odds are the ratio of the probability of success to the probability of failure (each understood to be a "proportion in the long run"); and in the subjective case, odds are the ratio of the amount a person *feels* he should get in return for risking a certain amount in a given situation. Important here is the word "feels," for different persons may well propose different gambling odds depending on their own personal evaluation of a situation. In contrast, when the concept of gambling odds is used in connection with the classical concept of probability or the frequency interpretation, it does not tell us how a person "feels"—it tells us how to bet on the basis of the possibilities pro and con, or on the basis of past experience.

In connection with subjective probabilities, we are usually more interested in converting odds to probabilities than probabilities to odds, for that is how subjective probabilities were defined. Nevertheless, if we want to convert a probability to odds, we have only to make use of the fact that since the probabilities for success and failure are, respectively, $p=\frac{a}{a+b}$ and $1-p=\frac{b}{a+b}$, their ratio is $\frac{p}{1-p}=\frac{a}{b}$. Thus, in the subjective case too, the odds for success are in the same ratio as the probability of success to the probability of failure.

EXERCISES

1. If data compiled by the manager of a department store show that 703 of 925 women who entered the store on a Saturday morning made at least one purchase, estimate the probability that a woman who enters the store on a Saturday morning will make at least one purchase.

2. The records of a life insurance company show that 1,436 of the 1,771 policy holders who took their policy out at age twenty-five were still alive at age fifty.

(a) Estimate the probability that a twenty-five-year-old person who takes out a policy with this company will still be alive at age fifty.

(b) Can we use the same figures to estimate the probability that *any* twenty-five-year-old person will still be alive at age fifty? Explain.

3. In a sample of 200 cans of mixed nuts (taken from a large shipment), 36 contained no pecans. Estimate the probability that there will be no pecans in a can of mixed nuts which is randomly selected from this shipment.

4. If 189 of 300 television viewers in a certain area expressed the opinion that the coverage of local news was inadequate, estimate the probability that a television viewer randomly selected in this area will feel that local news coverage is inadequate.

5. Weather bureau statistics show that in a certain community it has rained 24 times in the last 60 years on the first Sunday in June, the day which a service club has chosen for its annual picnic.

(a) What is the probability that it will rain on the day which the service club has chosen for its picnic?

(b) What are the odds that it will rain on that day?

(c) What are the odds that it will not rain on that day?

(d) If somebody offered us a bet of a quarter against our dime that it will *not* rain on the day of the service club's picnic, whom would this bet favor?

6. Among the 114 times that a race driver has raced his car, he won 38 times.

(a) What is the probability that he will win his next race?

(b) What are the odds that he will win his next race?

(c) If somebody offered us *even money* (odds of 1 to 1) that this race driver will win his next race, who would be favored by this bet?

7. In a sample of 856 cars stopped at a roadblock, the drivers of 214 cars did not have their seatbelts fastened.

(a) Estimate the probability that the driver of a car stopped at this roadblock will have his seatbelt fastened.

(b) What are the odds that the driver of a car stopped at this roadblock will have his seatbelt fastened?
(c) What are the odds that the driver of a car stopped at this roadblock will not have his seatbelt fastened?
(d) If we offered a friend a bet of our \$10 against his \$4 that the driver of the next car stopped at the roadblock will have his seatbelt fastened, who would be favored by this bet?

8. In a sample of 360 students attending a large university, 54 favored the continuation of a certain physical education requirement.
(a) Estimate the probability that if we randomly select one of the students attending this university, he will be against the continuation of the physical education requirement.
(b) What are the odds that a student thus selected will favor the continuation of the physical education requirement?
(c) What are the odds that a student thus selected will be against the continuation of the physical education requirement?
(d) If somebody offered to bet \$12 against our \$2 that a student thus selected will be against the continuation of the physical education requirement, who would be favored by this bet?

9. If a high school senior feels that 5 to 3 are fair odds that he will be admitted to the University of Southern California, what is his personal probability that this will be the case?

10. A sportswriter feels that the odds are 4 to 1 that the home team will lose an upcoming football game. What subjective probability expresses his feelings about the team winning or tieing that game?

11. One insurance salesman offers another insurance salesman a bet of \$55 against \$25 that he will not be able to sell a policy to a difficult client. If the second insurance salesman feels that this is fair, what subjective probability is he thus assigning to his selling a policy to this client?

12. Mr. Brown bet Mrs. Brown \$14 to \$2 that she will get a speeding ticket on their cross-country vacation. If Mrs. Brown considers this a fair bet, what personal probability does she assign to her getting a speeding ticket on that trip?

13. If the owner of a racehorse is willing to bet \$70 against \$20 that his horse will not win, what can we say about the probability which he assigns to this unfortuitious event? (*Hint*: the answer should read "at least...".)

14. If a stockbroker is unwilling to bet \$50 against \$150 that the price of a certain stock will go up within a week, what does this tell us about his subjective probability that this will be the case? (*Hint*: the answer should read "less than . . .".)

15. With reference to the illustration on page 52, suppose that the student is willing to bet \$5 against \$1 that he will get an A in the test, but not \$6 against \$1. What does this tell us about the subjective probability which he assigns to his getting an A in the test?

16. A doctor offers to bet a friend \$8 against \$2, but not \$13 against \$2, that he will get an emergency call during the round of golf they have planned for that afternoon. What does this tell us about the subjective probability he assigns to his getting such a call?

17. Mr. Jones is willing to bet Mrs. Jones \$15 to her \$5, but not \$8 to her \$2, that they will be late getting to a party. What does this tell us about the subjective probability he assigns to their getting there late?

18. A television executive is willing to bet \$1,000 against \$9,000, but not \$1,200 against \$8,800, that a new comedy show will be a success. What does this tell us about the executive's personal probability that the show will be a success?

19. Explain the fallacy of the following assertion: Since probabilities are measures of uncertainty, the probability we assign to an outcome will always increase when we get more information.

20. Discuss the following assertion: When a businessman claims that the chances of his making a profit that year are 40 percent, whatever happens that year cannot prove him right or wrong.

21. Explain how one might assign a probability to the truth of testimony given at a trial, using
(a) the frequency interpretation;
(b) subjective probabilities.

22. Is it possible for two persons using the frequency interpretation to arrive at different probabilities regarding the potential success of a new novel? Explain.

23. Discuss the following statement: "Textbooks on probability are often written with the intention of making the rules of probability plausible to the reader. Therefore, certain presuppositions are not mentioned but are regarded as 'understood,' as it is assumed that this kind of presentation is the safest means of instilling belief in the rules."

24. The following illustrates how "common sense" can be misleading in connection with probabilities or odds: A box contains 100 beads, some red and some white. One bead will be drawn, and you are asked to call beforehand whether it is going to be red or white. At what odds would you be willing to bet on this gamble if

(a) you have no idea how many red beads and how many white beads there are in the box;

(b) you are told that the box contains 50 red beads and 50 white beads?

25. Discuss the following argument, which is another example that illustrates how "common sense" can be misleading in connection with probabilities or odds: "Among three indistinguishable boxes one contains two pennies, one contains a penny and a dime, and one contains two dimes. We randomly select one of these boxes and take out one of the coins without looking at the other. If the coin we get is a penny, we may well be inclined to say that there is a *fifty-fifty chance* that the other coin in the box is also a penny. After all, the penny must have come either from the box with the penny and the dime or the box with the two pennies, and, hence, the other coin is a dime in one case and a penny in the other."

26. To get a "feeling" for the Law of Large Numbers, actually flip a coin 100 times and after each fifth flip plot the accumulated proportion of heads as in Figure 2.2 on page 49. By at most how much do these proportions differ from $\frac{1}{2}$ after the first 50 flips?

27. Actually roll a die 300 times and after each 10 rolls plot the accumulated proportion of 6's on a graph like that of Figure 2.2. By at most how much do these proportions differ from $\frac{1}{6}$ after the first 200 rolls of the die?

28. Replacing each card before the next one is drawn, actually make 200 drawings from an ordinary deck of 52 playing cards and after each 10 drawings plot the accumulated proportion of *spades* on a graph like that of Figure 2.2. By at most how much do these proportions differ from $\frac{1}{4}$ after the first 100 drawings?

SUMMARY

It is quite possible that the reader may dislike, or sharply disagree with, some of the ideas which we have expressed in this chapter. This has to be expected, since the subject is, indeed, controversial. Also, the reader may feel that the three probability concepts are really not as disparate as we have made them seem—in fact, he may feel that in the different probability concepts we are really looking at the same "thing" from different angles, or perhaps through differently colored glasses. Whatever it is that all these probability concepts have in common may not be immediately apparent, but we shall try to give a partial answer, at least, in Chapter 5.

As we have already tried to point out, each of the probability concepts has its good sides and its bad sides. The main criticism against the classical concept is that it applies only when the possibilities are all equally likely, and this not only limits its applicability, but raises the question of how we can ever really *know* whether two or more possibilities are, in fact, equally likely. So far as the frequency interpretation is concerned, it is criticized mostly because such probabilities always refer to what has happened and will happen in many similar situations, but never directly to the particular event with which we happen to be concerned. Also, some critics feel that the frequency concept should be reserved for situations which are easily repeatable, such as the roll of a die or the flip of a coin. They feel that it would be too farfetched for example, to say that the probability is 0.998 that Nero actually played his fiddle while Rome burned, where the 0.998 probability is really a measure of the *reliability of our historical information*, and hence might apply also to the question whether Henry VIII really had six wives. Of course, there is no harm in using probabilities this way, so long as everyone knows what they are meant to say.

So far as subjective probabilities are concerned, their main weakness is that they always represent only one man's or one woman's opinion. This raises questions about their practical value, for it is meaningless to ask whether a subjective probability is right or wrong, good or bad, true or false (unless, of course, a person lies *intentionally* when he is asked questions about fair odds). All this suggests that there should be some way in which subjective probabilities can be "policed," namely, some way in which we can decide whether a person's subjective probabilities can be of any practical value to anyone but the person, himself. One such "control" will be discussed in Chapter 5, where we shall learn about a criterion for judging the *consistency* of subjective probabilities, but in most cases this is not enough. The whole question of whether we should or should not accept a person's subjective probabilities really boils down to what we know about the person, himself, his experience in evaluating chances, his past performance in making predictions, his familiarity with the circumstances, and none of this is purely a matter of mathematics.

Another argument against subjective probabilities is that throughout history the human mind has proven itself singularly inept at estimating odds. This is true, especially, in situations where such estimates are based on common sense or intuition. The following is a classical example: There are 40 persons at a party and someone offers *even money* (odds of 1 to 1) that at least two of the 40 persons have the same birthday. Since there are 365 days and only 40 persons, common sense suggests that it should be quite unlikely that any two of them have the same birthday, but if we followed this "advice" and bet that no two persons at the party have the same birthday, we would be at a great disadvantage. It can be shown (see, for example, the book by S. Goldberg listed in the Bibliography at the end of this book) that the *correct odds* are better than 8 to 1 that at least two of the 40 persons have the same birthday. Indeed, had there been 50 persons at the party, the odds would have been better than 32 to 1.

3

EXPECTATIONS

INTRODUCTION

When we are faced with uncertainties, our decisions are usually not based on probabilities alone; in most cases we must also know something about the consequences (profits, losses, penalties, or rewards) to which we are exposed. For instance, if we have to decide whether to buy a new car, knowing that our old car will soon require repairs is not enough—to make an intelligent decision we would also have to know how much these repairs would cost. To give another example, suppose that a building contractor has to decide whether to bid on a construction job which promises a profit of \$40,000 with a probability of 0.20, or a loss of \$9,000 (due to bad estimates, strikes, late delivery of materials, and so on) with a probability of 0.80. Clearly, the probability of his making a profit is not very high, but the amount he stands to gain is much greater than the amount he stands to lose. Both of these examples demonstrate the need for a method of combining probabilities and consequences, and it is for this purpose that we shall introduce the concept of a **mathematical expectation**.

MATHEMATICAL EXPECTATION

If an insurance agent tells us that a person aged twenty-two can expect to live 51 more years, this does not mean that he really "expects" the person to live until his or her seventy-third birthday and then drop dead on the next day. Similarly, if we read that in the United States a married woman can expect to have 2.4 children, a person can expect to eat 10.4 pounds of cheese and 324 eggs a year, and that a child in the age group from five to 14 can expect to visit his dentist 1.9 times a year, it must be obvious that we are not using the word "expect" in its colloquial sense. A woman cannot very well have 2.4 children, a child cannot visit his dentist 1.9 times, and it would be surprising, indeed, if we found somebody who has actually eaten 10.4 pounds of cheese and 324 eggs in a given year. So far as the first statement of this paragraph is concerned, some persons aged twenty-two will live another 12 years, some will live another 30 years, some will live another 64 years, . . ., and the life expectancy of "51 more years" will have to be interpreted as an *average,* namely, as a *mathematical expectation.*

Originally, the concept of a mathematical expectation arose in connection with games of chance, and in its simplest form it can be stated as follows:

> **If a person stands to win the amount a with the probability p (and nothing with the probability $1-p$), his mathematical expectation is given by the product $a \cdot p$.**

Thus, if 1,000 raffle tickets are issued for a cash prize of $a = \$500$, the probability for each ticket is $p = \frac{1}{1{,}000}$ and the mathematical expectation of a person who holds one of these raffle tickets is $a \cdot p = 500 \cdot \frac{1}{1{,}000} = \$0.50$. This can be interpreted as an average in two different ways: We can argue that the 1,000 raffle tickets *together* will win \$500, and hence $\frac{500}{1{,}000} = \$0.50$ per ticket, or we can argue that if the kind of raffle were repeated a great number of times, a holder of a ticket should lose 99.9 percent of the time, win \$500 a tenth of a percent of the time, and hence win on the average \$0.50. Similarly, if we stand to win \$4 if a balanced coin comes up heads (and nothing if it comes up tails), our mathematical expectation is $4 \cdot \frac{1}{2} = \$2.00$; clearly, if this "game" were repeated a great many times, we should win \$4 about half the time, nothing the other half of the time, and hence our winnings should average \$2.00.

In the first example of the preceding paragraph, \$0.50 would be a *fair price* for one of the raffle tickets, and in the second example, \$2.00 would be a *fair price* to pay for the privilege of playing the game. By "fair" we mean that *the amount we pay equals the amount we can expect to win,* and in this sense a mathematical expectation is sometimes referred to as the **fair price of the uncertainty** with which we happen to be concerned. To consider another example, suppose we want to insure our car, worth \$3,000, against theft. The uncertainty against which we want to insure ourselves is whether or not our car will be stolen, and the "fair price" of this risk would have to depend on the chances of a theft. If statistics show that these chances are 1 in 200, the mathematical expectation which measures the "fair price" of this uncertainty is $\frac{1}{200} \cdot 3{,}000 = \$15$, and this is what an insurance company calls the pure premium. Of course, the actual premium we have to pay would also include a sales commission and other additions to account for the insurance company's expenses as well as a profit. Most car owners gladly pay more than the "fair price" of the uncertainty of having their car stolen, since the extra amount they have to pay in addition to the pure premium does not hurt, whereas an uninsured loss of $3,000 would. Of course, the insurance company does not worry about the $3,000 it may have to pay for the theft of any given car—if it insures a sufficiently large number of cars, all that matters is that *on the average* its losses should be close to $15 per car.

To illustrate how the concept of a mathematical expectation can be made more general, let us change the raffle on page 61 so that there is also a second prize of $200 and a third prize of $50. Now we can argue that one of the 1,000 tickets will pay $500, another will pay $200, and a third will pay $50, so that altogether they will pay $750 or $0.75 per ticket; this is the mathematical expectation for each ticket. Alternately, if this raffle were repeated many times, a holder of a ticket should lose 99.7 percent of the time and win each of the three prizes 0.1 percent of the time. On the average, a holder of a ticket should thus win

$$0(0.997)+500(0.001)+200(0.001)+50(0.001) = \$0.75$$

which is the sum of the products obtained by multiplying each amount by the corresponding probability. Generalizing from this example, let us now make the following definition:

If the probabilities of winning (getting) the amounts $a_1, a_2, \ldots,$ or a_k, are respectively, $p_1, p_2, \ldots,$ and p_k, then the mathematical expectation E is given by

$$E = a_1 p_1 + a_2 p_2 + \ldots + a_k p_k$$

In words, each amount is multiplied by the corresponding probability, and the mathematical expectation E is the *sum* of the products thus obtained. So far as the a's are concerned, let us point out that they are *positive* when they represent profits, winnings, or gains (namely, amounts which a player stands to receive), and that they are *negative* when they represent losses, penalties, or deficits (namely, amounts which a player stands to lose). For instance, if we bet 25 cents on the flip of a coin (that is, we either win 25 cents or lose 25 cents depending on the outcome), the amounts a_1 and a_2 are 25 and -25, the corresponding probabilities are $p_1 = \frac{1}{2}$ and $p_2 = \frac{1}{2}$, and the mathematical expectation is

$$E = 25 \cdot \tfrac{1}{2} + (-25) \cdot \tfrac{1}{2} = 0$$

This is what the expectation should be in an **equitable game**, namely, in a game which does not favor either player.

Although we referred to the quantities $a_1, a_2, \ldots,$ and a_k as "amounts," they need not be *cash* winnings, losses, penalties, or rewards. When we said on page 60 that in the United States a married woman can expect to have 2.4 children, we are actually saying that 2.4 is the sum of the products obtained by multiplying the *a*'s, which, referring to the number of children, can be 0, 1, 2, 3, ..., by the corresponding probabilities that a married woman in the United States will have that many children. Similarly, when we said on page 60 that a child in the age group from five to 14 can expect to visit his dentist 1.9 times a year, we are actually saying that 1.9 is the sum of the products obtained by multiplying the *a*'s, which, referring to the number of visits, can be 0, 1, 2, 3, ..., by the corresponding probabilities that a child in the age group from five to 14 will make that many visits to his dentist. As we shall see later, the *a*'s can also be arbitrary units of utility, "appreciation," or "discomfort."

To give another example where we use the above formula for *E*, suppose that somebody is offered $24,500 for his house, and that he must make up his mind whether to accept this offer or turn it down with the hope of getting a better price. If a well-informed real estate salesman tells him that his chances of selling the house for $27,000, $25,500, $24,500, or $22,000 are, respectively, 0.12, 0.21, 0.38, and 0.29, he figures that he can expect to get

$$27{,}000(0.12)+25{,}500(0.21)+24{,}500(0.38)+22{,}000(0.29) = \$24{,}285$$

and unless he is a *confirmed optimist* (or derives pleasure from the gamble), it stands to reason that he should accept the offer of $24,500.

In the last example the *a*'s were again sums of money, but they are not in the following situation pertaining to the number of persons there are in cars entering a given national park. If it is known from past experience that the respective probabilities are

Number of persons	1	2	3	4	5	6
Probability	0.06	0.32	0.27	0.25	0.08	0.02

it follows that the expected number of persons per car is

$$1(0.06)+2(0.32)+3(0.27)+4(0.25)+5(0.08)+6(0.02) = 3.03$$

This kind of information may well come in handy in determining the adequacy of picnic facilities, camping facilities, and so on. For instance, if it is known that the park facilities can accommodate only 2,500 persons, the park administration will avoid problems by not admitting more than $\frac{2,500}{3.03}$ or just about 825 cars at any given time.

EXERCISES

1. If a service club sells 500 raffle tickets for a color television set worth $380, what is the mathematical expectation of a person who buys one of these tickets?
2. A person gets $2.60 if he draws an ace from an ordinary deck of 52 playing cards and nothing if he draws any other card. How much should he be charged for playing this game so that he can expect to break even?
3. Box A contains 20 slips of paper of which 19 are marked $0 and the other is marked $5; Box B contains 50 slips of paper of which 49 are marked $0 and the other is marked $13. If a person will receive whatever is on the slip which he draws, should he prefer to make a drawing from Box A or from Box B?
4. With reference to the raffle described on page 61 (where there was only one prize), what is the mathematical expectation of a person who has
   (a) bought two tickets;
   (b) bought three tickets?
5. To introduce his new cars to the public, a dealer offers a prize of $5,000 to the lucky person who comes to his showrooms, inspects the new models, and writes his comments as well as his name on a special card. What is each entrant's mathematical expectation, if 20,000 persons came to the dealer's showrooms, looked at the new models, and filed such cards? Of course, the winning card is presumably drawn at random. Also, was it worthwhile to spend 30 cents on gasoline to drive to this dealer's showrooms?
6. Repeat Exercise 5 with the modification that there is also a second prize of $2,000.
7. As part of a promotional scheme, the manufacturer of a new breakfast food offers a prize of $50,000 to someone willing to try the new product (distributed without charge) and send in his name on the label. The winner is to be drawn at random from all the entries in front of a large television audience.
   (a) What is each entrant's mathematical expectation, if 2,000,000 persons send in their names?
   (b) Was it worth the 8 cents postage it cost to send in one's name?

8. The two finalists in a golf tournament play 18 holes, with the winner getting \$15,000 and the runner-up getting \$9,000. What are the two players' mathematical expectations if
(a) they are evenly matched;
(b) the better player should be favored by odds of 2 to 1?

9. In a winner-take-all bowling tournament among four professional bowlers, the prize money is \$8,000. If one of these bowlers figures that the odds against his winning are 4 to 1, what is his mathematical expectation? Would he be better off if he made a secret agreement with the other bowlers to split the prize money evenly regardless of who wins?

10. If someone were to give us \$2.00 each time we roll a 6 with a balanced die, how much should we pay him when we roll a 1, 2, 3, 4, or 5 to make the gamble equitable?

11. If someone were to give us \$1.00 each time we roll a 1 with a balanced die, \$2.00 each time we roll a 2, and \$3.00 each time we roll a 3, how much should we pay him when we roll a 4, 5, or 6 to make the game equitable?

12. If the two teams are evenly matched, the probabilities that a baseball World Series will last four, five, six, or seven games, are respectively, $\frac{1}{8}$, $\frac{1}{4}$, $\frac{5}{16}$, and $\frac{5}{16}$ (as the reader will be able to verify later on).
(a) If the two teams are evenly matched, how many games can we expect a baseball World Series to last?
(b) If somebody has tickets for the sixth and seventh games, what are the probabilities that he will see zero, one, or two games, and how many games can he expect to see?

13. A student's parents promise him a gift of \$10 if he gets an A in mathematics, \$5 if he gets a B, and otherwise no reward. What is his expectation if the probability of his getting an A is 0.32 and the probability of his getting a B is 0.40?

14. Two friends are betting on repeated flips of a balanced coin. One has \$5 at the start and the other has \$3, and after each flip the loser pays the winner one dollar. If p is the probability that the one who starts with \$5 will win his friend's \$3 before he loses his own \$5, explain why $3p-5(1-p)$ should equal 0, and then solve the equation $3p-5(1-p)=0$ for p. Generalize this result to the case where two players start with a dollars and b dollars, respectively.

15. A grab-bag contains 20 packages worth 7 cents apiece, 20 packages worth 12 cents apiece, and five packages worth 14 cents apiece. Is it worthwhile to pay 8 cents for the privilege of choosing one of these packages at random? Is it worthwhile to pay 10 cents? Is it worthwhile to pay 12 cents?

16. A nurseryman has to decide whether to bid on the landscaping of a public building. What should he do if he figures that the job promises a profit of \$5,400 with a probability of 0.40, or a loss (due to a lack of rain or perhaps an early frost) of \$3,500 with a probability of 0.60?

17. If it is extremely cold in the Midwest, a resort in Arizona will have 240 guests during the Christmas season; if it is cold (but not extremely cold) in the Midwest they will have 207 guests; and if the weather is moderate in the Midwest they will have only 132 guests. How many guests can they expect if the probabilities for extremely cold, cold, or moderate weather in the Midwest are, respectively, 0.34, 0.58, and 0.08?

18. At a refreshment stand at the beach, the probabilities that a customer will order a hamburger for 40 cents, a cheeseburger for 50 cents, or a hot dog for 30 cents are, respectively, 0.58, 0.23, and 0.19. Also, the probabilities that he or she will buy a soft drink for 25 cents or a cup of coffee for 20 cents are, respectively, 0.52 and 0.35, and the probability that he or she will buy an order of french fries for 30 cents is 0.28. How much can a customer be expected to spend at this refreshment stand if he does not order more than one sandwich, one drink, and one order of french fries?

19. An importer is offered a shipment of transistor radios for \$5,000, and the probabilities that he will be able to sell them for \$6,000, \$5,500, \$5,000, or \$4,500 are, respectively, 0.25, 0.46, 0.19, and 0.10. If he buys the shipment, what is his expected gross profit?

20. A wage negotiator of a labor union feels that the odds are 3 to 1 that the members of the union will get a raise of 50 cents in their hourly wage, the odds are 17 to 3 against their getting a raise of 25 cents in their hourly wage, and the odds are 9 to 1 against their getting no raise at all. What is the corresponding expected raise in their hourly wage?

21. The following table gives the probabilities that a woman who enters a given department store will make zero, one, two, three, or four purchases:

Number of purchases	0	1	2	3	4
Probability	0.26	0.33	0.28	0.09	0.04

How many purchases can a woman entering this department store be expected to make?

22. A fire chief knows that the probabilities for zero, one, two, three, four, or five fires on any given day are as shown in the following table:

Number of fires	0	1	2	3	4	5
Probability	0.22	0.34	0.25	0.13	0.05	0.01

How many fires can he expect per day? (It is assumed here that the probability of more than five fires is negligible.)

EXPECTATIONS AND SUBJECTIVE PROBABILITIES

In all of the problems about mathematical expectations in the preceding section we were given the values of a and p (or the values of the a's and p's) and we were asked to calculate the corresponding expectations. In this section we shall consider problems in which we are given the values of E and a, so that we can use the equation $E = a \cdot p$ to solve for the probability p. In the next section we shall consider problems in which we are given the values of E and p, so that we can use the equation $E = a \cdot p$ to solve for the amount a about which the uncertainty exists.

Although we presented a fairly straightforward method of evaluating subjective probabilities in Chapter 2, it should be noted that this method also has some serious shortcomings. For one thing, if the amount of money involved is very small, most persons tend to be rather careless in judging odds; also, some persons object to any form of gambling on moral grounds, while others, for one reason or another, refuse to risk any part of their capital regardless of the odds. Of course, we cannot force anyone to gamble with his own money, but there is a way around this, which is illustrated in the following example: Suppose that a playwright stands to get \$25,000 if his new play is a success, nothing if it flops, but he has the option of settling now for \$15,000 in lieu of all future claims. What we would like to know is the value of the subjective probability p which he assigns to the play's success. Although we may not be able to determine the exact value of p, we can learn a lot about it from his decision whether or not to accept the cash settlement of \$15,000. If he *rejects* it, we can argue that \$15,000 is less than his expectation of $a \cdot p = 25{,}000 \cdot p$, namely, that $15{,}000 < 25{,}000p$ and, hence, $p > \frac{15{,}000}{25{,}000}$ (or $\frac{3}{5}$). Correspondingly, if he *accepts* the cash settlement, $15{,}000 \geqq 25{,}000p$ and, hence, $p \leqq \frac{15{,}000}{25{,}000}$ (or $\frac{3}{5}$). Similarly, as the reader will be asked to verify in Exercise 1 on page 71, we would find that $\frac{3}{5} < p \leqq \frac{4}{5}$, if he *rejected* a cash settlement of \$15,000 but would be willing to accept a cash settlement of \$20,000. To narrow it down further, we might ask the playwright what he considers a "fair" settlement, and if his answer is \$18,000, we could equate this amount to $a \cdot p = 25{,}000p$, getting $18{,}000 = 25{,}000p$ and $p = \frac{18}{25}$ (or 0.72).

The \$18,000 which the playwright is thus willing to accept may well be referred to as his **subjective expectation**—it is the mathematical expectation $a \cdot p$, with p being the subjective probability he assigns to the uncertainty in question. Thus, if we know a person's subjective expectation E with regard to an uncertainty and the value of a, we can determine the corresponding subjective probability p by solving the equation $E = a \cdot p$.

In many cases, subjective expectations can be determined by simply asking the person what he considers a "fair price" of the uncertainty or gamble; if this does not produce an honest answer, we can always force the issue by determining (on a put-up or shut-up basis) whether he will accept or reject specific cash alternatives in lieu of the uncertainty and thus get an interval for p, as in the playwright example on page 67.

Note that this method of measuring subjective probabilities makes it possible to avoid the problems which arise when a person is unwilling to risk any part of his *own* capital, that is, when he is unwilling to gamble with his *own* money regardless of the odds. *All we require of a person is that he will not refuse an outright gift in exchange for a small mental effort*. To illustrate, let us refer again to the example on page 52, where we discussed the problem of measuring the probability which a student assigns to his getting an A in a final examination. Suppose now that we offer this student a reward of \$2.00 if he gets the A, with the alternative of settling for the *certainty* of receiving a somewhat smaller amount regardless of what grade he will get in the examination. If he tells us that he is "pretty sure" that he will get an A, but that he would consider it fair and reasonable to settle now for \$1.50 cash, we can equate this amount to his expectation $\$2.00 \cdot p$, getting $1.50 = 2.00p$ and hence $p = \frac{3}{4}$. This does not agree with the result which we obtained on page 52, where we showed that p is on the interval from $\frac{4}{5}$ to $\frac{7}{8}$, but it should really not come as a surprise that *we may get different results when we measure something in two different ways*. This is true, particularly, when the measurements involve an element of subjectivity—clearly, one English teacher may feel that an essay is worth an A while another English teacher may feel that it is worth only a B, and one art critic may feel that a painting is superb, while another critic may feel that it is mediocre, at best.

The two methods of measuring subjective probabilities which we have introduced here and in Chapter 2 *are different*, and in practice they may or may not produce different results. This is true even though *theoretically* the results should be the same. To illustrate, let us refer again to the businessman who feels that a bet of \$500 against \$200 that his new bookstore will make a profit during the first year is *fair*. At the time, we used these odds to determine the corresponding probability (namely, the businessman's subjective probability that the store will make a profit during the first year) as $p = \frac{500}{500+200} = \frac{5}{7}$; now let us use these betting odds to determine the businessman's expectation. Since he stands to win \$200 with the probability p or lose \$500 with the probability $1-p$, his expectation is

$$200p + (-500)(1-p) = 700p - 500$$

and since the bet is supposed to be fair, this expectation must be zero. Thus, we get $700p - 500 = 0$ and $p = \frac{500}{700} = \frac{5}{7}$, and it can be seen that the two methods have yielded the same result. The same argument holds in general. If a person feels that a to b are fair odds for a "success" and p is the probability (his personal probability) of success, he feels, in other words, that his expectation

$$b \cdot p + (-a) \cdot (1-p)$$

should equal zero. If we solve the resulting equation for p, we get

$$bp - a + ap = 0$$

$$(b+a)p = a$$

and finally $p = \frac{a}{a+b}$, which is identical with the formula on page 51. *This shows that, although we now have two different ways of measuring subjective probabilities (asking for fair odds or working with expectation), they are theoretically, or conceptually, the same.*

UTILITY

Since the value which we attach to things is a personal matter, it is difficult to measure. Of course, we can always ask a person *directly* how much something is worth, but "words are cheap" and unless there is something at stake it is hard to get straight answers. Suppose, for instance, that a friend who is a sports enthusiast tells us that he would "give his right arm" for a ticket to a championship football game which has been sold out for weeks. Naturally, this is only a figure of speech, but it is meant to imply that the value which our friend attaches to such a ticket is very high. If he were asked to be more specific, he might say that he would be willing to pay \$25 or even \$100, but unless we can put this on a "put-up or shut-up" basis, it really does not have much significance. Thus, suppose that we propose to him the following deal: For \$5.00 we will let him draw one of 20 sealed envelopes, of which 19 contain a dollar bill and the other contains a ticket to the football game. If he accepts this deal and we let U denote the "cash value" which he assigns to a ticket to the game, we can argue that his expectation, namely, $\frac{1}{20}\cdot U+\frac{19}{20}\cdot 1$, must be *at least* \$5.00. Symbolically,

$$\tfrac{1}{20}\cdot U+\tfrac{19}{20}\cdot 1 \geqq \$5.00$$

and this leads to $U+19 \geqq 5\cdot 20$ upon multiplying by 20, and hence to $U \geqq \$81$. Thus, we have arrived at the result that our friend feels that the ticket to the game is worth at least \$81, and if we varied the odds

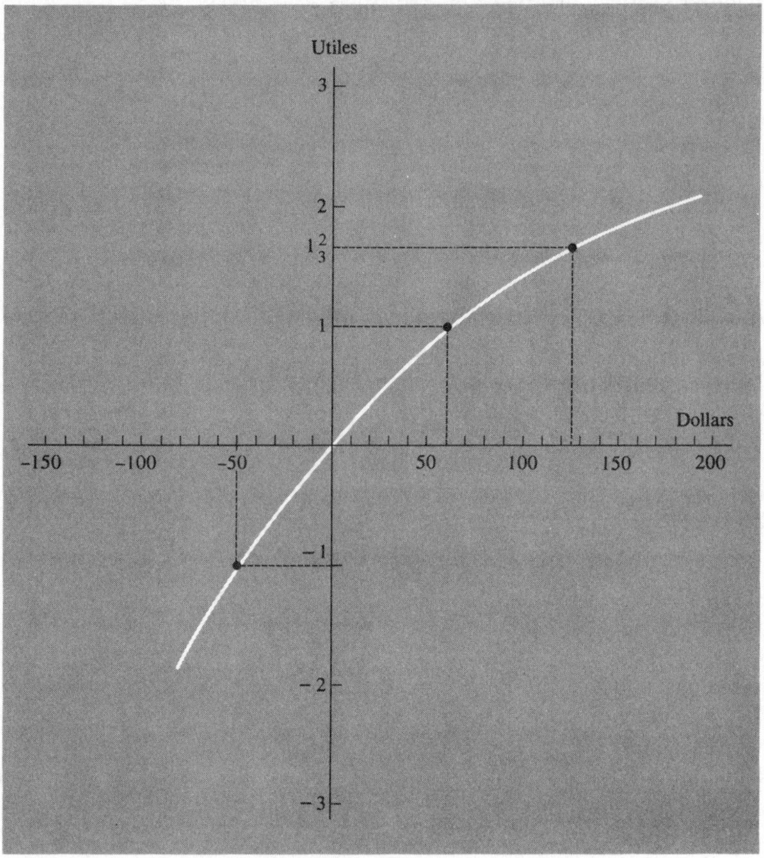

FIGURE 3.1 A utility curve.

and the amounts (the number of envelopes and the amounts which they contain), we could narrow it down more than that. This, incidentally, is the example which we referred to on page 67, when we said that we would study problems in which we are given E and p, and solve for a.

It was assumed in this example that \$5 is worth five times \$1 and \$100 is worth twenty times \$5, and assumptions like this are generally reasonable so long as we are dealing with small amounts. It is a well-known fact, however, that *the utility which a person assigns to a sum of money does not always equal its monetary value.* Very often, the reason for this is that "the second \$100 is worth less than the first," "the second \$1,000 is worth less than the first," ..., or that "the second \$1,000 a person *owes* hurts more than the first." Of course, \$100 is \$100 and \$1,000 is \$1,000, but the *value* of a sum of money will depend on how much a person already has or owes; in technical language, this is the problem of **marginal utility.**

To give another example, suppose that we offer someone the choice between an outright gift of \$60 or a gamble on the flip of a coin which pays \$125 if called right and nothing if called wrong. Even though the mathematical expectation of the gamble is \$62.50 and exceeds the outright gift, we should be surprised if a person actually selected the gamble. In some cases this may be due to superstitions or an aversion to gambling, but in most cases it is simply due to the fact that the first \$60 is worth more to a person than the next \$65.

To carry this a bit further, suppose that the gamble is modified so that the person wins the \$125 if he draws a white bead out of an urn containing white beads and black beads. If he feels that the value of the gamble just about equals the outright gift of \$60 when there are 60 white beads and 40 black beads, we can write

$$\frac{60}{60+40}\cdot U = \$60$$

where U is the value, or utility, which the person attaches to \$125. This leads to $U = \frac{5}{3}\cdot\$60$, but it is better not to multiply this out and get $\frac{5}{3}\cdot 60 = \$100$, since the value which a person attaches to \$100 need not be $\frac{5}{3}$ times the utility he assigns to \$60. To avoid this kind of problem, it is best to introduce an artificial unit of utility and call it a **utile**. For instance, in our example we could have let \$0 and \$60 be 0 and 1 utiles, respectively, and we could then say that the person assigns \$125 a utility of $\frac{5}{3}\cdot 1 = 1\frac{2}{3}$ utiles.

Now suppose that we offer the same person another proposition—we offer to bet him \$60 on the flip of a balanced coin. Of course, to make the bet fair he should really pay us \$60 when we win, but he may feel that this is not worth the risk—to him, losing \$60 may do more harm than winning \$60 does good, and he may be willing to put up only \$50 against our \$60. Of course, we would not take this bet, but it tells us something about the utility which the person assigns to a loss of \$50. If we write this utility as U, we find that

$$\tfrac{1}{2}\cdot U + \tfrac{1}{2}\cdot 1 = 0$$

making use of the fact that we already assigned the person's winning \$60 a value of 1 utile. Then, if we solve this equation, which arose from the fact that the person considered it "fair" to bet only \$50 against our \$60, we arrive at the result that $U = -1$, namely, that the utility which he assigns to losing \$50 is -1 utiles. Continuing in this way, we could find the utilities, number of utiles, which the person assigns to various other amounts, and thus arrive at a **utility curve** like the one shown in Figure 3.1.

EXERCISES

1. With reference to the example on page 67, show that if the playwright rejected a cash settlement of \$15,000 but accepted \$20,000, this would lead to the result that $\frac{3}{5} < p \leqq \frac{4}{5}$, where p is the probability he assigns to the play's success.

2. A recent college graduate is faced with a decision which cannot wait; he must decide whether to accept or turn down a job paying \$8,600 a year. What can we say about the probability which he assigns to his only other prospect, a job paying \$12,900 a year, if he decides to take the \$8,600 job?

3. An insurance company agrees to pay the promoter of a drag race \$6,000 in case the event has to be cancelled because of rain. If the company's actuary feels that a fair net premium for this risk would be \$960, what probability does he assign to the possibility that the race may have to be cancelled because of rain?

4. Mr. Brown has the option of keeping a gift of \$5.00 or betting it on the outcome of a baseball game. If the gamble will pay him \$15 if and only if the home team wins and *he cannot make up his mind whether to keep the money or bet it on the game,* what does this tell us about the probability which he assigns to the home team winning the game?

5. To handle a liability suit, a lawyer has to decide whether to charge a straight fee of \$400 or a contingent fee of \$1,200 which he will get only if his client wins. What does the lawyer think about his client's chances if

(a) he prefers the straight fee of \$400;
(b) he prefers to take the gamble;
(c) he cannot make up his mind?

6. A baseball pitcher has to choose between a straight salary of \$30,000 and a salary of \$20,000 with a bonus of \$40,000 if he wins twenty games. How does this pitcher feel about his chances of winning twenty games if

(a) he decides to accept the straight salary of \$30,000;
(b) he prefers the \$20,000 salary with the possibility of the bonus;
(c) he cannot make up his mind?

7. One contractor offers to do a road repair job for \$12,000, while another contractor offers to do it for \$15,000 with a penalty of \$5,000 if the job is not finished on time. How does the person who lets out the contract for the job feel about the chances that the second contractor will not be able to complete the job on time if

(a) he gives the job to the first contractor;
(b) he gives the job to the second contractor;
(c) he canot make up his mind?

8. The manufacturer of a new battery additive has to decide whether to sell his product for \$0.80 a can, or for \$1.20 with a "double-your-money-back-if-not-satisfied" guaranty. How does he feel about the chances that a person will ask for double his money back if

(a) he decides to sell the product for \$0.80;
(b) he decides to sell the product for \$1.20 with the guaranty;
(c) he cannot make up his mind?

9. Mr. Smith feels that it is about a toss-up whether to accept \$13.00 cash or to gamble on drawing a bead from an urn containing 25 white beads and 75 red beads, where he is to receive \$2.50 if he draws a white bead and a new electric drill if he draws a red bead. What value, or utility, does he attach to the new electric drill?

10. Mr. Black has the choice of staying home and watching television, to which he assigns a utility of 15 utiles. If, instead, he goes to a party, he may have a terrible time, to which he assigns a utility of -20 utiles, or a wonderful time, to which he assigns a utility of 50 utiles. What does he think of his chances of having a good time at the party if

(a) he decides to stay home and watch television;
(b) he decides to go to the party;
(c) he cannot make up his mind?

11. With reference to the example on page 70, what can we say about the utility our friend assigns to a ticket to the football game, if
(a) he refused to pay \$8.00 for drawing one of the envelopes;
(b) he is willing to pay \$6.00 but not \$7.50 for the gamble;
(c) he cannot make up his mind whether to pay \$7.20 for the gamble?

12. Suppose that the person referred to on page 71, who attaches a utility of 1 utile to \$60 is willing to pay \$60, but not more than \$60, for a gamble which will pay him \$300 if he draws a spade from an ordinary deck of 52 playing cards, but nothing if he draws a heart, diamond, or club. What utility does he assign to \$300?

13. Suppose that the person referred to on page 71, who attaches a utility of -1 utiles to a loss of \$50, actually owes a friend \$50. To cancel the debt, he proposes the following scheme, which he considers "fair": If a balanced die comes up 1, 2, 3, 4, or 5 the debt is cancelled, but if the die comes up 6 he will owe his friend \$150. What utility does he assign to a loss of \$150?

14. Read the necessary values of the graph of Figure 3.1 to find the utilities (in utiles) which the person in question assigns to
(a) making a profit of \$150;
(b) having a loss of \$100;
(c) making a profit of \$200;
(d) having a loss of \$25.

15. The person whose utility curve is given in Figure 3.1 takes on a job which promises to pay him \$50, \$100, \$150, or \$200 with respective probabilities of 0.24, 0.32, 0.28, and 0.16. What is his expected utility?

16. Discuss under what conditions it would be smart for a person to have
(a) complete health insurance;
(b) only major medical insurance which does not cover, say, the first \$500;
(c) no health insurance whatsoever.

DECISION MAKING

When faced with uncertainties, we can often use mathematical expectations in making decisions. In the example on page 63, for instance, we suggested that the person should accept the \$24,500 offer for his house, since otherwise his expectation would be only \$24,285, which is less. In general, if we have to choose between two or more alternatives, it is considered "rational" to take the one with the "most promising" mathematical expectation, namely, the one which *maximizes expected gains, minimizes expected costs, maximizes one's expected utility, minimizes expected taxes,* and so on. Although this approach to decision making has great intuitive appeal, it is not without complications. In many problems it is difficult, if not impossible, to assign values to all of the a's (amounts) and p's (probabilities) in the formula for E on page 62. This is true especially when the p's are interpreted subjectively and represent only one man's opinion, or when the a's are given in units of utility, or perhaps units of satisfaction or appreciation, and not in strictly monetary terms.

To illustrate some of these difficulties, let us study the following situation: Mr. Green, the manager of the new products division of a tire manufacturer, knows that a competitor will introduce a fairly expensive new non-skid tread design, and he must decide whether his company should do the same. He figures that if he recommends that the new tread design be put on the market and it gains public acceptance, this will be worth \$2,000,000 to his company; on the other hand, if his company puts the new tread design on the market and it does not gain public acceptance, the resulting loss will be \$1,200,000. Also, if he recommends that the new tread design *not* be put on the market and the one introduced by the other company gains public acceptance, this will entail a loss of \$440,000 to his company (largely, for being put at a competitive disadvantage); and finally, if he recommends that the new tread design *not* be put on the market and the one introduced by the other company does not gain public acceptance, this is worth \$20,000 to his company (for having made a better decision than its competitor). Schematically, all this information can be summarized as follows:

	The new tire gains public acceptance	*The new tire does not gain public acceptance*
The company puts the new tire on the market	\$2,000,000	−\$1,200,000
The company does not put the new tire on the market	−\$440,000	\$20,000

Evidently, it will be advantageous to market the new tire only if it gains public acceptance, and Mr. Green's decision will therefore have to depend on the chances that this will be the case. Suppose, for instance, he feels that there is a fifty-fifty chance that the public will be willing to pay for the more expensive non-skid tread. He can then argue that *if his company puts the new tire on the market,* its *expected gain* is

$$2{,}000{,}000 \cdot \tfrac{1}{2} + (-1{,}200{,}000) \cdot \tfrac{1}{2} = \$400{,}000$$

and *if his company does not put the new tire on the market,* its *expected gain* is

$$(-440{,}000) \cdot \tfrac{1}{2} + 20{,}000 \cdot \tfrac{1}{2} = -\$210{,}000$$

Since an expected gain of \$400,000 is obviously preferable to an *expected loss* (negative gain) of \$210,000, it stands to reason that Mr. Green should not hesitate in recommending that the new tire be put on the market.

Observe, however, that the conclusion at which we have arrived is based on Mr. Green's appraisal of the chances for the new tire's success. What if his judgment is poor? What if his immediate superior feels that the odds against public acceptance of the new tire are, say, 3 to 1? In that case, the company can expect to gain

$$2{,}000{,}000 \cdot \tfrac{1}{4} + (-1{,}200{,}000) \cdot \tfrac{3}{4} = -\$400{,}000$$

if the new tire *is* put on the market, and

$$(-440{,}000) \cdot \tfrac{1}{4} + 20{,}000 \cdot \tfrac{3}{4} = -\$95{,}000$$

if the new tire is *not* put on the market. Choosing the lesser of two evils, that is, preferring an expected loss of \$95,000 to an expected loss of \$400,000, Mr. Green's superior would thus overrule his recommendation. All this serves to illustrate that mathematical expectations should be used as a tool in decision making only if one's judgment in assessing probabilities or odds is good.

The way in which we have studied this problem is referred to as a **Bayesian analysis**. In this kind of analysis, probabilities are assigned to the possibilities about which uncertainties exist (the so-called "States of Nature," which in our example were the commercial success or failure of the new tire); then, we choose the alternative which promises the greatest expected profit or the smallest expected loss. Some books dealing in detail with this kind of analysis are listed in the Bibliography at the end of the book.

Let us now examine briefly what Mr. Green might have done if he had no idea about the new tires' chances of gaining public acceptance, for in that case he could not have based his decision on mathematical expectations. To suggest one possibility, suppose that Mr. Green is a *confirmed optimist.* Looking at the situation through rose-colored glasses and always hoping for the best, he finds that his company may gain as much as \$2,000,000 if the new tire is put on the market; otherwise, it can at best gain

\$20,000. If it is for this reason that he recommends that the new tire be put on the market, he is, in fact, *maximizing the company's maximum gain*; that is, he is choosing the alternative for which the company's greatest gain is a maximum.

Now suppose that Mr. Green is a *confirmed pessimist*. Always looking for the worst that can possibly happen, he finds that his company may lose \$1,200,000 if the new tire is put on the market; otherwise, it cannot do worse than lose \$440,000. If it is for this reason that he recommends that the new tire not be put on the market, he is, in fact, *minimizing the company's maximum losses,* and we say that he is using the **minimax criterion**.

There are several other ways in which decisions can be made in the absence of any knowledge about the probabilities of the various "States of Nature." For instance, if Mr. Green is the kind of person who is always *afraid to lose out on a good deal,* he might argue that if he decides to put the tire on the market and it does not gain public acceptance, his company would have been better off by

$$20{,}000-(-1{,}200{,}000)=\$1{,}220{,}000$$

if he had decided not to put the tire on the market. Similarly, if he decides not to put the tire on the market and the one introduced by the competitor gains public acceptance, his company would have been better off by

$$2{,}000{,}000-(-440{,}000)=\$2{,}440{,}000$$

if he had decided to put the tire on the market. These differences are sometimes referred to as **opportunity losses** (or **regrets**), and the whole situation can be pictured as follows:

	The new tire gains public acceptance	*The new tire does not gain public acceptance*
The company puts the new tire on the market	0	\$1,220,000
The company does not put the new tire on the market	\$2,440,000	0

To explain the two 0's, note that when the company puts the new tire on the market and it turns out that it gains public acceptance, the decision was the right one and there is no loss of opportunity. The same is true also when the company does not put the new tire on the market and it turns out that the one introduced by the competitor does not gain public acceptance. Now, if Mr. Green is the kind of person who always wants to hold his opportunity losses to a minimum (or as we said above, does not want to lose out on a good deal), he would probably choose the alternative for which the greatest opportunity loss is a minimum and recommend that the new tire be put on the market. This is another use of the *minimax criterion* referred to on page 77 (which is now applied to the opportunity losses instead of the original "payoffs").

There are also situations in which the criteria which we have discussed are outweighed by other considerations. For instance, if Mr. Green of our example knows that he will be fired if his company loses more than $500,000 as a result of his decision, he would be foolish (though perhaps unselfish) to recommend that the tire be put on the market. The situation would be reversed if he knows that he will be fired unless his next major decision leads to a *substantial* gain. Further examples of situations where extraneous factors affect decisions are given in Exercises 3 and 15 on pages 81 and 85; it is hoped that they will amplify the point we have been trying to make that *there is no universal rule or criterion which will always lead to the best possible decision.*

Finally, suppose that Mr. Green simply cannot make up his mind and in desperation decides to flip a coin—*heads* he will recommend that the new tire be put on the market and *tails* he will recommend that the new tire not be put on the market. More or less as a figure of speech, we might say that he is thus leaving the decision "in the hands of the Gods." We said "more or less" for throughout history simple games of chance have been used to communicate with, or seek guidance from, the supernatural. In some primitive societies a person's innocence or guilt was determined by drawing lots, in ancient Greece and Rome oracles based predictions on casts of *astragali* (forerunners of modern dice), and in the Bible there is a reference to an occasion where the direction in which an army was to proceed was determined by shaking arrows in a quiver and observing the direction in which the first one fell. Nowadays, this kind of *divination* is frowned upon in most circles, but we still hear people appealing to "Lady Luck," while others pull petals out of daisies counting "She loves me, she loves me not, she loves me, she loves me not, . . ." Indeed, it was less than 250 years ago that John Wesley, the founder of Methodism, sought divine guidance with regard to a possible marriage by fasting, praying, and then drawing one of three lots advising him, respectively, to "Marry," "Think not of it this year," and "Think of it no more."

Looking at the situation mathematically, we find that in the last example the odds were fixed at 2 to 1 against an immediate marriage. Also, if Mr.

Green actually decided to base his decision on the flip of a coin, this would automatically make it an even bet that his company will put the new tire on the market. Of course, if Mr. Green chose to base his decision on the roll of a die with the provision that the new tire be put on the market unless the die comes up 6, he would be fixing the odds at 5 to 1 in favor of this alternative. Thus, even when we want to leave things to chance, there still remains the problem of how to fix (some might say "rig") the odds.

Let us investigate briefly what Mr. Green is actually doing when he bases his decision on the flip of a coin or the roll of a die. In the first case, where he uses the coin, we find that the company can expect to gain

$$2{,}000{,}000 \cdot \tfrac{1}{2} + (-440{,}000) \cdot \tfrac{1}{2} = \$780{,}000$$

if the new tire gains public acceptance, and

$$(-1{,}200{,}000) \cdot \tfrac{1}{2} + 20{,}000 \cdot \tfrac{1}{2} = -\$590{,}000$$

if the new tire does not gain public acceptance. On the other hand, if he uses a die as suggested, we find that the company can expect to gain

$$2{,}000{,}000 \cdot \tfrac{5}{6} + (-440{,}000) \cdot \tfrac{1}{6} = \$1{,}593{,}333$$

if the new tire gains public acceptance, and

$$(-1{,}200{,}000) \cdot \tfrac{5}{6} + 20{,}000 \cdot \tfrac{1}{6} = -\$996{,}667$$

if the new tire does not gain public acceptance. We are now faced with a situation which is just as bad as the one with which we began on page 78—using the die with 5 to 1 odds favoring the first alternative has potentially a greater expectation than using the coin *if it turns out that the new tire will gain public acceptance.* The situation is reversed *if it turns out that the new tire will not gain public acceptance.*

If Mr. Green chose the *minimax criterion* to select one of the two gambles (namely, select the one for which the *greatest expected loss is the least*), he would end up using the coin, but this raises the question whether there might be some other gambling scheme which is preferable even to the use of the coin. The answer is "Yes," but we shall not discuss it at this time. For one thing, Mr. Green would probably get fired if it were discovered that he is so "ignorant" that he has to base his decisions on the flip of a coin or the roll of a die. Also, the minimax criterion reflects sheer pessimism, and there is really no reason for such a bleak outlook in the given situation.

In the beginning of this section we pointed out that it is sometimes difficult to base decisions on mathematical expectations, as this requires knowledge of all the a's (amounts) and p's (probabilities) in the formula for E. We have already seen how a decision can be reversed by a change in the probabilities, and in Exercise 4 on page 81 the reader will be asked to check whether this can also happen when there are changes in the a's. In our example in the text we did not question the figures in the table on page 75, but in most practical situations the problem of assigning "cash values" to the consequences of one's decisions can pose very serious difficulties. There can be no doubt that the methods of this section provide useful tools for making decisions, but it should always be remembered that their application requires a great deal of care and discretion.

EXERCISES

1. A group of businessmen is faced with the problem whether to put up the funds needed to build a new sports arena or continue to hold its sports promotions in a college gymnasium. They figure that if the new arena is built and they can get a professional basketball franchise, there will be a profit of $2,050,000 during the next five years; if the new arena is built and they cannot get a professional basketball franchise, there will be a deficit of $500,000 during the next five years; if the new arena is not built and they get a professional basketball franchise, they will make a profit of $1,000,000 during the next five years; and if the new arena is not built and they cannot get a professional basketball franchise, there will be a profit of only $100,000 during the next five years (from their other promotions).

(a) Present all this information in a table like that on page 75.

(b) If they believe an official of the professional basketball organization who tells them that the odds are 2 to 1 *against* their getting the franchise, what should the group of businessmen decide so as to maximize the expected profit for the next five years?

(c) If they believe the sports editor of a local newspaper who tells them that the odds are only 3 to 2 *against* their getting the franchise, what should the group of businessmen decide so as to maximize the expected profit for the next five years?

(d) If one of these businessmen is a confirmed pessimist, would he be inclined to vote for or against putting up the funds for the new arena? Explain your answer.

(e) If one of these businessmen is a confirmed optimist, would he be inclined to vote for or against putting up the funds for the new arena? Explain your answer.

(f) Construct a table like that on page 77, which shows the opportunity losses (regrets) associated with the various possibilities. What action should the group of businessmen take if they wanted to hold the greatest possible opportunity loss to a minimum?

(g) Referring to the odds of part (b) and the opportunity losses of part (f), what action should the businessmen take so as to minimize their *expected* loss of opportunity?

2. With reference to Exercise 1, suppose that the businessmen reach their decision by flipping a coin. What is their expected profit for the next five years if

(a) they get the professional basketball franchise;

(b) they do not get the franchise?

3. With reference to Exercise 1, what would one expect the businessmen to do if

(a) they will be out of business unless they can make a profit of at least $1,200,000 during the next five years;

(b) they will be out of business unless they make a profit of at least $50,000 during the next five years.

4. Referring to the illustration in the text, suppose that the accounting department of the tire manufacturer discovers the following mistake: If the new tire is put on the market and gains public acceptance, the company stands to gain $4,000,000 instead of $2,000,000. Would this affect Mr. Green's decision when he feels that

(a) there is a fifty-fifty chance that the new tire will gain public acceptance;

(b) his superior is right and the odds are 3 to 1 that the new tire will *not* gain public acceptance?

5. Referring to the illustration in the text, find the company's expected gains corresponding to each of Mr. Green's decisions, when

(a) the odds against public acceptance of the new tire are 7 to 5;

(b) the odds against public acceptance of the new tire are 7 to 3;

(c) the odds against public acceptance of the new tire are 2 to 1.

Also state in each case which decision will maximize the company's expected gain.

6. Tom is starting out to meet his friends at the beach, but he cannot remember whether he was supposed to meet them in La Jolla or in Mission Beach, which are four miles apart. He lives 11 miles from the spot where he would meet them in La Jolla and nine miles from the spot where he would meet them in Mission Beach.

(a) Construct a table (similar to the one on page 75) which shows the number of miles Tom has to drive to meet his friends depending on where he goes first and where they are actually supposed to meet.

(b) Where should he go first if he wants to minimize the distance he can expect to drive to meet his friends and he feels that the odds are 5 to 1 that they are in La Jolla?

(c) Where should he go first if he wants to minimize the distance he can expect to drive to meet his friends and he feels that the odds are 2 to 1 that they are in La Jolla?

(d) Show that it does not matter where he goes first when the odds are 3 to 1 that his friends are in La Jolla.

(e) Where should he go if he were a confirmed pessimist?

(f) Where should he go if he were a confirmed optimist?

(g) Where should he go if he wants to minimize the greatest possible regret?

7. With reference to Exercise 6, suppose that Tom makes his decision by drawing a card from an ordinary deck of 52 playing cards and going to Mission Beach first and if and only if it is a spade. How many miles can he expect to drive to meet his friends if

(a) they are in La Jolla;

(b) they are in Mission Beach?

8. With reference to Exercise 6, show that if Tom bases his decision on the flip of a balanced coin, the number of miles he can expect to drive to meet his friends is the same regardless of whether they are in La Jolla or in Mission Beach.

9. Mrs. Jones, who lives across the Bay, plans to spend an afternoon shopping in San Francisco, and she has some difficulty deciding whether or not to take along her raincoat. If it rains, she will be inconvenienced if she does not bring it along, and if it does not rain, she will be inconvenienced unnecessarily if she does. On the other hand, it will be convenient to have the raincoat if it rains, and we shall say she is neither convenienced nor inconvenienced if she does not bring her raincoat and it does not rain. To express all this numerically, suppose that the numbers in the following table are *units of inconvenience*, so that the negative values reflect *convenience*:

	It rains	*It does not rain*
She takes the raincoat along	-10	6
She does not take the raincoat along	20	0

(a) What should Mrs. Jones do to *minimize her expected inconvenience* if she feels that the odds against rain are 6 to 1?
(b) What should Mrs. Jones do to *minimize her expected inconvenience* if she feels that the odds against rain are 3 to 1?
(c) Does it matter whether or not Mrs. Jones takes along her raincoat if she feels that the odds against rain are 5 to 1?
(d) What should Mrs. Jones do in order to minimize the greatest possible loss of opportunity?

10. With reference to Exercise 9, suppose that Mrs. Jones bases her decision on the flip of a coin. What inconvenience can she expect if (a) it rains; (b) it does not rain?

11. With reference to Exercise 9, suppose that Mrs. Jones numbers nine slips of paper from 1 to 9, and will take the raincoat if and only if the number she draws is less than 6. What inconvenience can she expect if (a) it rains; (b) it does not rain? Also compare this gamble with the one of Exercise 10, and judge which is "better" by comparing the respective maximum inconveniences.

12. With reference to the illustration in the text, suppose that Mr. Green uses a gambling scheme which will make him decide to put the new tire on the market with the probability p, and not to put the new tire on the market with the probability $1-p$. Express in terms of p the company's expected gains corresponding to whether or not the tire will gain public acceptance, and show that they are equal when $p=\frac{23}{183}$. Also show that with regard to the minimax criterion on page 77, this gambling scheme is preferable to the ones where he bases his decision on the flip of a coin or the roll of a die.

13. With reference to the illustration in the text, suppose that Mr. Green has the alternative of putting the tires on the market on a trial basis at a cost of \$100,000, after which he will be *certain* (for all practical purposes) whether the new tire will gain public acceptance. Of course, this raises the question whether it is worthwhile to spend the \$100,000. To answer it, let us take Mr. Green's original evaluation of the chances that the new tire will gain public acceptance, namely, fifty-fifty. If Mr. Green knew *for sure* whether the new tire will gain public acceptance, the right decision would yield his company either \$2,000,000 or \$20,000 (the corresponding entries in the table on page 75), and since the corresponding probabilities are supposedly $\frac{1}{2}$ and $\frac{1}{2}$, we find that the company can thus expect to gain

$$2{,}000{,}000\cdot\tfrac{1}{2}+20{,}000\cdot\tfrac{1}{2}=\$1{,}010{,}000$$

and this is what is called the **expected value of perfect information**. *In general, this is the amount one can expect to gain, profit, or win in a given* *situation if one always makes the right decision.* Since the value which we obtained here exceeds by $610,000 the $400,000 which the company can expect to gain if the tire is put on the market immediately (see page 76) and by $1,220,000 the $210,000 which the company can expect to lose if the tire is not put on the market, it stands to reason that the $100,000 would be well spent.

(a) With reference to Exercise 1 and the odds of part (b), what is the expected value of perfect information? Would it be worthwhile to the businessmen to spend $50,000 to find out for sure whether or not they will get the franchise before they make their decision about the arena?

(b) With reference to Exercise 6 and the odds of part (c), what is the expected mileage which corresponds to perfect information. If Tom figures that it costs him 3 cents per mile to drive his car, would it be worthwhile for him to spend 10 cents on a phone call to find out whether his friends are in La Jolla or in Mission Beach?

14. A dinner guest wants to show his appreciation to his hostess by sending her either a dozen roses or a pound of candy. He remembers, though, that she is either allergic to roses or on a strict reducing diet, but he can't remember which. In any case, he feels that her reaction to the gift will be as shown in the following table, where the numbers are in *units of appreciation*:

	Hostess is allergic to roses	*Hostess is on strict diet*
He sends roses	−5	12
He sends candy	3	−10

(a) What should the dinner guest send so as to maximize the *expected appreciation* of his gift, if he feels that the odds are 4 to 1 that the hostess is allergic to roses rather than dieting?

(b) What should the dinner guest send so as to maximize the *expected appreciation* of his gift, if he feels that the odds are 2 to 1 that the hostess is allergic to roses rather than dieting?

(c) With reference to the odds of part (b), what is the expected value of perfect information? Assuming that each unit of appreciation is equivalent to $0.80, would it be worth a $1.80 long-distance phone call to find out whether she is allergic to roses or on a diet?

(d) What should the dinner guest do if he wanted to minimize the greatest possible regret?

(e) Suppose that the dinner guest considers basing his decision on the draw of a card from an ordinary deck of 52 playing cards, sending the roses if he draws a 10, jack, queen, king, or ace, and sending the candy if he draws a 2, 3, 4, 5, 6, 7, 8, or 9. Check whether it would be better in that case not to send any gift at all.

15. With reference to Exercise 14, what should the dinner guest do if he is told that the hostess never invites anyone again

(a) if she receives a "thank you" gift worth less than 5 negative units of appreciation;

(b) unless she receives a "thank you" gift worth at least 5 units of appreciation.

4

EVENTS

INTRODUCTION

In the study of probability it is convenient to represent the various possibilities by means of numbers or points, for we can then talk *mathematically* about anything that may or may not happen without having to go through lengthy verbal details. Actually, this is what we do in sports when we refer to players by their numbers, it is what the Internal Revenue Service does when it refers to taxpayers by their social security numbers, and it is precisely what we do when we use ZIP codes to represent various locations. In this chapter we shall examine briefly some of the problems that arise when we identify the various things that can happen in a given situation by means of numbers, points, sets of numbers, or sets of points.

THE SAMPLE SPACE

In many branches of mathematics we refer to the set which consists of all the things with which we are concerned as the *universe of discourse,* or simply the *universal set*. Here, we shall borrow from the language of statistics and refer to the set of all the outcomes that are possible in a given situation as the corresponding **sample space.** For instance, in Case A on page 33 the sample space consisted of the 20 ways in which the teacher can select one of the 20 novels; in the example on page 22 the same space consisted of the 220 ways in which a housewife can choose three of the 12 kinds of cheese; and in Exercise 18 on page 13 the sample space consisted of the 6,561 ways in which a person can answer the questionnaire.

The use of points rather than numbers has the added advantage that it makes it easier to *visualize* the various possibilities and perhaps discover some of the special features which different outcomes have in common. To give an example, suppose that as part of a survey on popular music we ask students whether they like a certain new record, dislike it, or don't care. The possible replies of *one* student are pictured in Figure 4.1, where we represented them by means of three points, to which we assigned the

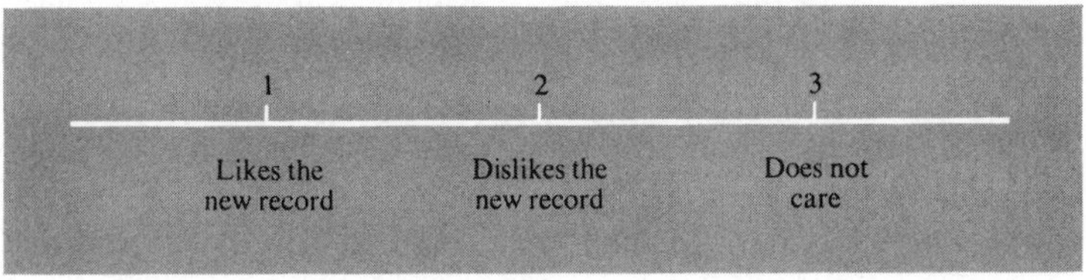

FIGURE 4.1 Sample space showing responses of one student.

(code) numbers 1, 2, and 3. Actually, these points could have been drawn in any pattern, not necessarily on a straight line, and we could have assigned to them any arbitrary set of numbers, say 0, 19, and 6, or -2, 21, and -7.

If we now consider the reactions of two students, we find that there are

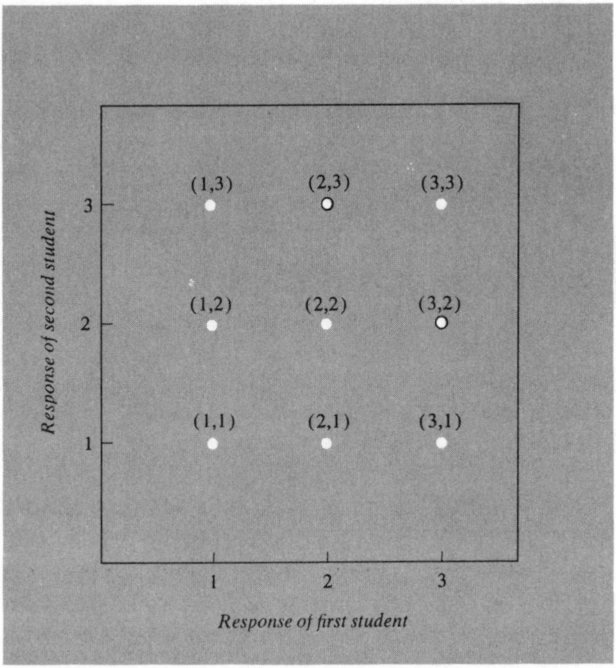

FIGURE 4.2 Sample space showing responses of two students.

nine possibilities, which we can represent by means of nine points as in Figure 4.2. Again, we could have drawn these points in any arbitrary pattern (perhaps, along a circle or the sides of a triangle), but our choice of the particular two-dimensional arrangement shown in Figure 4.2 has the advantage that the points can easily be identified. If we use the same coding as before, with 1, 2, and 3 standing for "like the record," "dislike it," and "don't care," we can now use the *coordinates* of each point to indicate the reactions of the two students. For instance, the point (2, 1) represents the case where the first student dislikes the new record and the second student likes it, and the point (1, 3) represents the case where the first student likes the new record while the second student doesn't care, the point (3, 2) represents the case where the first student doesn't care and the second student dislikes the new record, and the point (2, 2) represents the case where both students dislike the new record.

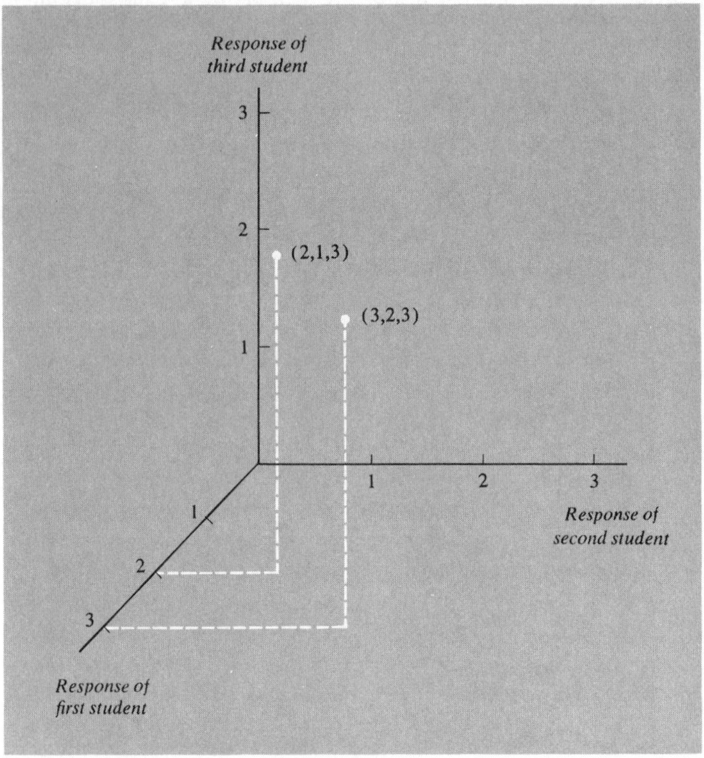

FIGURE 4.3 Two points of the sample space for responses of three students.

An important feature of the notation which we have used in this example is that it can easily be generalized. For instance, for the reactions of three students we could let (2, 1, 3) represent the case where the first student dislikes the new record, the second student likes it, and the third student doesn't care; similarly, we could let (3, 2, 3) represent the case where the second student dislikes the new record and the first and third students don't care. These points can be pictured as in Figure 4.3, but our power of visualization would fail if we had to consider the responses of four students, or more. For four students we could still let (2, 1, 3, 1) represent the case where the second and fourth students like the new record, the first student dislikes it, and the third student doesn't care, but we cannot picture these points as in Figures 4.1, 4.2, and 4.3, for this would require a space of *four dimensions.* Of course, we could identify all the possibilities with an arbitrary set of points, say, along a circle, but it is doubtful whether this would be of any value.

Sample spaces are often classified according to the number of dimensions in which the points are arranged, and in the preceeding example it seemed natural to use one dimension for the responses of one student, two dimensions for the responses of two students, and three dimensions for the responses of three students. Nevertheless, it should be observed that the dimensions of the sample space may differ, depending on what we consider distinct possibilities, or distinct outcomes. Suppose, for example, that in the two-student case we are interested only in *how many* of the students like the new record, *how many* dislike it, and *how many* do not care. In that case it would have been preferable to use the three-dimensional sample space of Figure 4.4 instead of the two-dimensional sample space of Figure 4.2. In Figure 4.4 the first coordinate gives the number of students who like the new record, the second coordinate gives the number of students who dislike it, and the third coordinate gives the number of students who don't care. Thus, the point (0, 2, 0) represents the case where both students interviewed dislike the new record, and the point (1, 0, 1) represents the case where one of the two students likes the new record while the other doesn't care. Note that the points (2, 3) and (3, 2) circled in Figure 4.2 *together* represent the same outcome as the point (0, 1, 1) circled in Figure 4.4; that is, when one of the two students dislikes the new record and the other does not care there are actually *two possibilities,* between which we distinguish in the sample space of Figure 4.2 but not in the sample space of Figure 4.4. Also, the points (1, 3) and (3, 1) of Figure 4.2 *together* represent the same outcome as the point (1, 0, 1) of Figure 4.4, and the point (1, 1) of Figure 4.2 *by itself* represents the same outcome as the point (2, 0, 0) of Figure 4.4.

Generally, it is desirable to use sample spaces whose points represent **primitive outcomes** in the sense that they do not consist of two or more

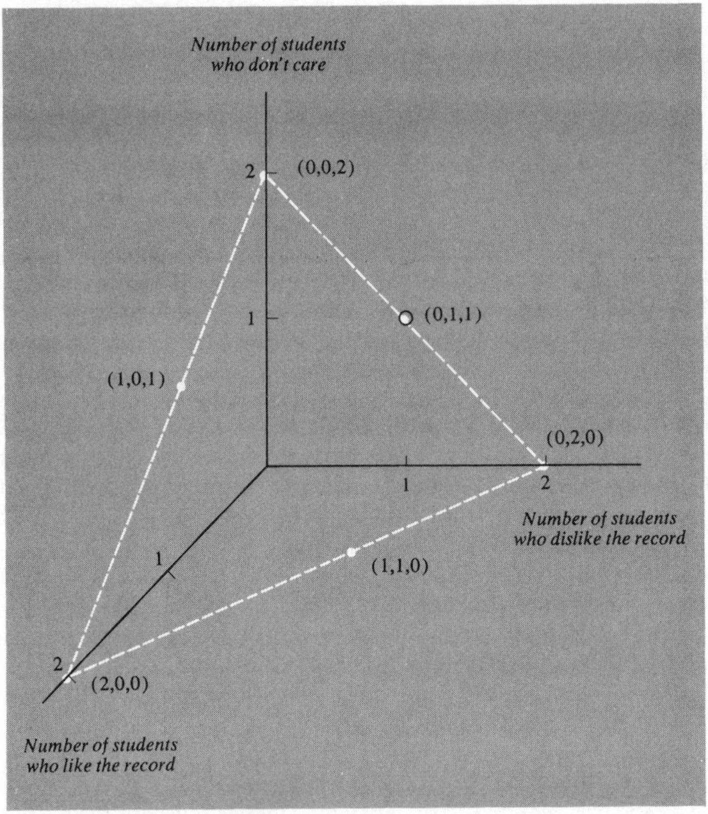

FIGURE 4.4 Sample space for responses of two students.

possibilities which are distinguishable in some fashion. Unless there is a special reason for not adhering to this rule, we would thus use the sample space of Figure 4.2 in preference to that of Figure 4.4.

Although it may be helpful to know the number of dimensions in which we plot the points of a sample space, it is more common to classify sample spaces according to the number of points which they contain. All of the samples spaces mentioned so far have been **finite**; that is, they consisted of a finite, or fixed, number of points. Other examples of finite, though much larger, sample spaces are the one representing the 16,777,216 ways in which one can answer a multiple-choice test of 12 questions with each question having four possible answers, or the one which represents the 24,040,016 five-man committees which can be selected from among a company's 80 employees (see pages 10 and 23). In this chapter and the next we shall consider only sample spaces which are finite, but to give the reader some idea about a sample space that is **infinite,** suppose that a coin is flipped until *tails* appears for the first time. Unless the coin is two-headed, the chances are that this will happen fairly soon, but so far as the sample space is concerned, we have to consider the possibilities that it will happen on the first flip of the coin, on the second flip, the third, the fourth, the fifth, and so on. Clearly, there is no limit to the number of possibilities, for if we stopped at any given number, say, 10,000 it is conceivable (though highly unlikely) that we will get at least 10,000 heads before we get the first tails, and hence have to go beyond 10,000 flips. Infinite sample spaces arise also when we are dealing with measurements (say, of temperatures or weights) which take on values on a continuous scale. This case will not be treated in this book, but the one where there are as many possibilities as there are positive integers will be discussed briefly in Chapter 7.

EVENTS

In Chapter 2 we assigned probabilities to outcomes or possibilities, namely, to anything that may or may not take place; from now on we shall refer to all such happenings as **events**. Actually, *"event" is the non-technical term which corresponds to "subset of a sample space" in the language of mathematics,* and in case the reader is not familiar with the term "subset," let us point out that a subset is any part of a set, including the set as a whole and, trivially, the **empty set** $\varnothing$, which has no elements at all. For instance, in Figure 4.2 the subset which consists only of the point $(1, 2)$ represents the *event* that the first student interviewed likes the new record while the second student does not; also, the subset which consists of the points $(1, 1)$, $(2, 2)$, and $(3, 3)$ corresponds to the *event* that both students respond in the same way, and the subset which consists of the points $(1, 3)$, $(2, 3)$, $(3, 3)$, $(3, 1)$, and $(3, 2)$ represents the *event* that at least one of the two students doesn't care.

Still referring to the two-student-interview example and the sample space of Figure 4.2, suppose that we let A denote the event that the first student likes the new record, B denote the event that the second student dislikes the record, C denote the event that each of the two students either likes or dislikes the record, and D denote the event that one of the two students likes the new record while the other doesn't care.* Referring to Figure 4.5, which

* It is customary in mathematics to denote sets with capital letters.

shows the same sample space as Figure 4.2, we find that event A is represented *geometrically* by the three points inside the solid line, event B is represented by the three points inside the dashed line, event C is represented by the four points inside the dotted line, and event D consists of the two points which are circled in the diagram.

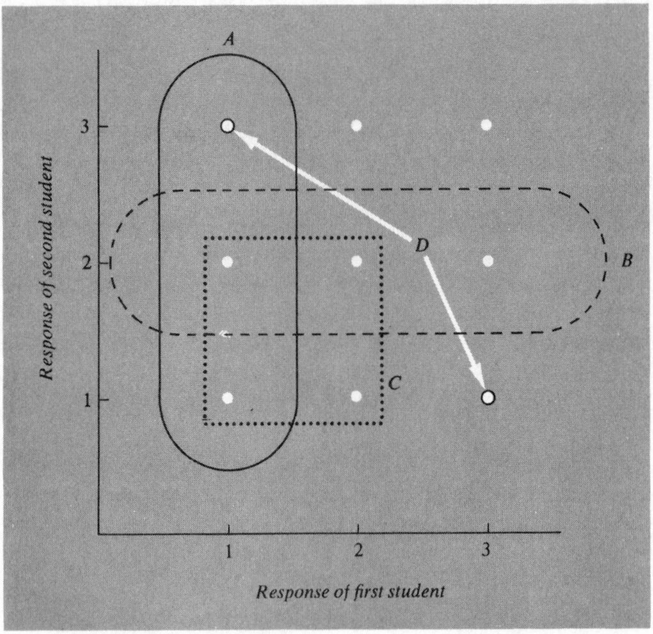

FIGURE 4.5 Sample space showing responses of two students.

Algebraically, we can write $A = \{(1, 1), (1, 2), (1, 3)\}$, which in the usual notation of sets means that A has as its elements (that is, consists of) the three points $(1, 1)$, $(1, 2)$, and $(1, 3)$. Similarly, $B = \{(1, 2), (2, 2), (3, 2)\}$, $C = \{(1, 1), (1, 2), (2, 1), (2, 2)\}$, and $D = \{(1, 3), (3, 1)\}$. Note that events B and D have no points in common—this means that they cannot *both* occur, and we refer to them as **mutually exclusive**. As can easily be seen, events C and D are also mutually exclusive, but events A and B are not.

To consider another example, suppose that on a Friday morning a baker bakes four chocolate cream pies, which have to be sold on that Friday or Saturday, or they will spoil. Using two coordinates so that $(1, 2)$, for example, represents the event that he sells one on Friday and two on

Saturday, (4, 0) represents the event that he sells all four on Friday, (1, 1) represents the event that he sells one on Friday and one on Saturday, we find that there are altogether 15 possibilities, as pictured in Figure 4.6. Furthermore, if R is the event that he sells all of the pies, we find that

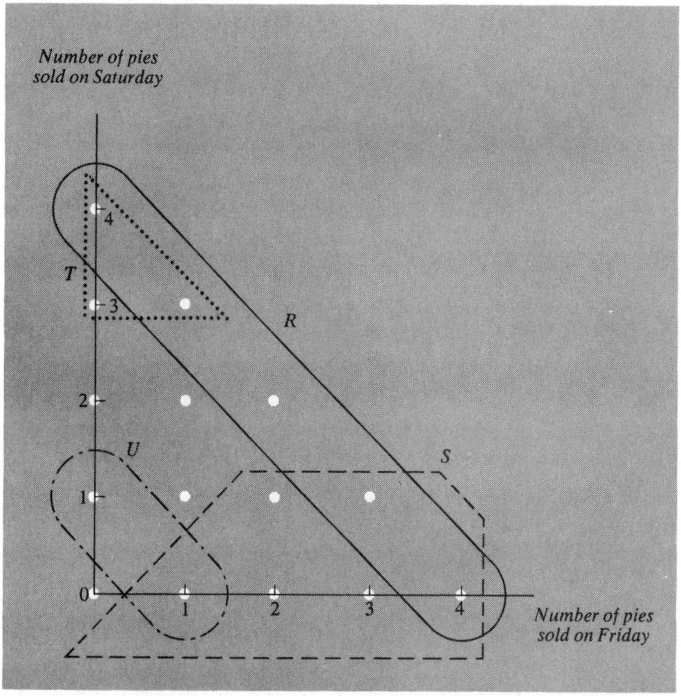

FIGURE 4.6 Sample space showing sales of pies.

$R = \{(0,4), (1,3), (2,2), (3,1), (4,0)\}$, and if S, T, and U are, respectively, the events that he sells more pies on Friday than on Saturday, that he sells at least three pies on Saturday, and that he sells only one pie, we find that $S = \{(1,0), (2,0), (2,1), (3,0), (3,1), (4,0)\}$, $T = \{(0,3), (1,3), (0,4)\}$, and $U = \{(0,1), (1,0)\}$. In Figure 4.6, the points belonging to R are inside the solid line, those belonging to S are inside the dashed line, those belonging to T are inside the dotted line, and those belonging to U are inside the line which is alternately dotted and dashed. Note that S and T, R and U, and T and U are mutually exclusive, while the other three pairs, R and S, R and T, and S and U, are not.

EXERCISES

1. Suppose that in a true-false test we let 1 indicate that the answer to a question is right and 0 that it is wrong. Furthermore, we shall write (1,0), for example, to indicate that the answer to the first of two questions is right while the other is wrong, and (1,1,0) to indicate that the answers to the first two of three questions are right while the third answer is wrong.

(a) Draw a sample space similar to that of Figure 4.1, showing the two ways in which one question can be answered.

(b) Draw a sample space similar to that of Figure 4.2, showing the four ways in which two questions can be answered right or wrong. Indicate the points as well as their coordinates.

(c) Draw a sample space similar to that of Figure 4.3, showing the eight ways in which three questions can be answered right or wrong. Indicate the points as well as their coordinates.

(d) List the points of the sample space of part (b) which correspond to the event that one answer is right and one answer is wrong.

(e) List the points of the sample space of part (c) which correspond to the event that two answers are right and one answer is wrong.

(f) List the points of the sample space of part (c) which correspond to the event that at least two of the three answers are right.

(g) Describe *in words* what event is represented by (1,0,0,1).

(h) List the points of the sample space for the answers to four true-false questions which correspond to the event that two answers are right and two answers are wrong.

(i) Describe *in words* what event is represented by (0,0,0,1,1).

(j) List the four points of the sample space for the answers to five true-false questions which correspond to the event that the first three answers are right.

(k) If we are interested only in the total number of correct answers, how many points would there be in the respective sample spaces which we would use instead of those of parts (a), (b), and (c).

2. In Figure 4.3 we indicated only two of the 27 points of the sample space for the responses of three students who are asked about the new record. With reference to this sample space list the points which together represent

(a) the event that all three respond in the same way;

(b) the event that the first and second students both dislike the new record;

(c) the event that the first student likes the record while the other two either both dislike it or both do not care;

(d) the event that two and only two of the students don't care;

(e) the event that none of the three students likes the new record.

3. On page 89 we indicated that the points (2, 3) and (3, 2) of Figure 4.2 together represent the same event as the point (0, 1, 1) of Figure 4.4. Which point or points of Figure 4.2 correspond to
(a) the point (0, 2, 0) of Figure 4.4;
(b) the point (0, 0, 2) of Figure 4.4;
(c) the point (1, 1, 0) of Figure 4.4?

4. With reference to Figure 4.2 list the points which together represent the event that
(a) the first student doesn't care;
(b) at least one of the two students likes the new record;
(c) either both students dislike the record or they both don't care;
(d) the two students respond differently.

5. With reference to Figure 4.6, list the points which together represent the event that
(a) the baker will sell as many of the pies on Friday as on Saturday;
(b) the baker will sell two of the pies on Friday;
(c) the baker will sell one more pie on Friday than on Saturday;
(d) altogether the baker will sell at most two of the pies.

6. With reference to Figure 4.6, express *in words* what events are represented by the following subsets:
(a) $V = \{(0,0), (0,1), (0,2), (0,3), (0,4)\}$;
(b) $W = \{(1,1), (1,2), (1,3), (2,1), (2,2), (3,1)\}$;
(c) $X = \{(2,0), (2,1), (2,2), (3,0), (3,1), (4,0)\}$?

7. With reference to Exercise 6, which of the following pairs of events are mutually exclusive: (a) V and W; (b) V and X; and (c) W and X?

8. A small taxicab company owns four cabs. Using two coordinates so that (3, 1), for example, indicates that three of the cabs are in operative condition of which one is out on a call, and (2, 2) indicates that two of the cabs are in operative condition and they are both out on a call, draw a diagram (similar to that of Figure 4.2) which shows the 15 points of the corresponding sample space. Also, if K is the event that at least two of the cabs are out on a call, L is the event that exactly one of the operative cabs is not out on a call, and M is the event that only one of the cabs is operative, list
(a) the points which comprise event K;
(b) the points which comprise event L;
(c) the points which comprise event M.
Also indicate these three events on the diagram by enclosing the points which comprise events K, L, and M, respectively, with a solid line, a dashed line, and a dotted line.

9. With reference to the sample space of Exercise 8, express *in words* what events are represented by the following sets of points:
(a) $N = \{(1, 1), (2, 1), (3, 1), (4, 1)\}$;
(b) $O = \{(2,0), (3,0), (3, 1), (4,0), (4, 1), (4, 2)\}$;
(c) $P = \{(0, 0), (1,0), (1, 1)\}$.

10. With reference to Exercises 8 and 9, which of the following pairs of events are mutually exclusive:
(a) K and M;
(b) K and O;
(c) M and N;
(d) L and P;
(e) O and P;
(f) K and N?

11. In a small restaurant, guests are served in two dining rooms, of which the smaller has three tables and the larger has four tables. Using two coordinates so that (2, 3), for example, represents the event that at a given moment two tables in the small room and three tables in the large room are being used, and (1,0) represents the event that at a given moment one table in the small room and none of the tables in the large room are being used, draw a diagram (similar to that of Figure 4.2) which shows the 20 points of the corresponding sample space. Also, if Q is the event that at least five of the tables in the restaurant are being used, R is the event that the same number of tables is being used in both rooms, and S is the event that fewer tables are being used in the large room than in the small room, list the points which
(a) comprise event Q;
(b) comprise event R;
(c) comprise event S.
Also indicate events Q, R, and S on the diagram by enclosing the points which they comprise with a solid line, a dashed line, and a dotted line.

12. With reference to Exercise 11, express *in words* what events are represented by the following sets of points:
(a) $T = \{(0, 4), (1, 3), (2, 2), (3, 1)\}$;
(b) $U = \{(3, 4), (2, 4), (1, 4), (0, 4)\}$;
(c) $V = \{(0, 2), (1, 3), (2, 4)\}$;
(d) $W = \{(0, 0), (0, 1), (0, 2), (0, 3), (0, 4), (1, 0), (2, 0), (3, 0)\}$.

13. With reference to Exercises 11 and 12, which of the following pairs of events are mutually exclusive:
(a) Q and S;
(b) R and U;
(c) T and W;
(d) U and V;
(e) Q and W;
(f) T and R?

14. Among the applicants for a teaching position in an elementary school in Scottsdale, Arizona, Miss A holds an M.A. degree and is not a native of Arizona, Mrs. B does not hold an M.A. degree and is not a native of Arizona, Mrs. C. holds an M.A. degree and is a native of Arizona, Miss D does not hold an M.A. degree but is a native of Arizona, Mrs. E holds an M.A. degree and is a native of Arizona, and Mrs. F does not hold an M.A. degree but is a native of Arizona. One of these six applicants is to get the job, and the event that it is given to someone with an M.A. degree, for example, is denoted $\{A, C, E\}$. Indicate in a similar manner the events that the job is given to

(a) a person who calls herself Mrs.;
(b) a native of Arizona without an M.A. degree;
(c) a single person who is not a native of Arizona;
(d) someone who either holds an M.A. degree or who is married but not a native of Arizona.

15. Among the eight cars which a dealer has in his showroom, Car 1 is new, has airconditioning, power steering, and bucket seats; Car 2 is one year old, has airconditioning, but neither power steering nor bucket seats; Car 3 is two years old, has airconditioning and power steering, but no bucket seats; Car 4 is three years old, has airconditioning, but neither power steering nor bucket seats; Car 5 is new, has no airconditioning, no power steering, and no bucket seats; Car 6 is one year old, has power steering, but neither airconditioning nor bucket seats; Car 7 is two years old, has no airconditioning, no power steering, and no bucket seats; and Car 8 is three years old, has no airconditioning, but power steering as well as bucket seats. If a customer buys one of these cars, and the event that he chooses a new car, for example, is represented by the set {Car 1, Car 5}, indicate similarly the sets which represent the events that he chooses

(a) a car without airconditioning;
(b) a car without power steering;
(c) a car with bucket seats;
(d) a car that is either two or three years old.

16. Which of the following pairs of events are mutually exclusive? Explain your answers.

(a) Having rain and sunshine on Easter Sunday, 1977.
(b) A person leaving Los Angeles by jet at 11 P.M. and arriving in Chicago on the same day.
(c) Being under thirty years of age and being president of the United States.
(d) A driver getting a ticket for speeding and his getting a ticket for going through a red light.

(e) A person wearing a red shirt and a blue tie.
(f) Getting a king and a queen when drawing one card from an ordinary deck of 52 playing cards.
(g) Getting a king and a black card when drawing a card from an ordinary deck of 52 playing cards.
(h) A baseball player getting a walk and hitting a home run in the same game.
(i) A baseball player getting a walk and hitting a home run in the same at bat.
(j) A baseball player hitting the ball inside the ballpark and getting a home run in the same at bat.

COMPOUND EVENTS

There are many situations in which we are interested in events that are actually combinations of several simpler kinds of events. For instance, in the example on page 93 we might be interested in the event that *either the baker sells all the pies or he sells at least three on Saturday,* the event that *he sells only one pie and he sells more on Friday than on Saturday,* or the event that *he does not sell all the pies.* In the first case we are interested in the event that *either* R *or* T *occurs,* which is represented by the points $(0,4)$, $(1,3)$, $(2,2)$, $(3,1)$, $(4,0)$, and $(0,3)$; in the second case we are interested in the event that U and S *both occur,* which is represented by the single point $(1,0)$; and in the third case we are interested in the event that R *does not occur,* which is represented by the points $(0,0)$, $(0,1)$, $(0,2)$, $(0,3)$, $(1,0)$, $(1,1)$, $(1,2)$, $(2,0)$, $(2,1)$, and $(3,0)$.

To simplify expressions like these and avoid ambiguities, it will be convenient to use here the customary notation of sets. Thus, we shall write the event that *either* R *or* T *occurs* as $R \cup T$, and refer to it as the **union** of R and T. In case the reader is not familiar with the language of sets, let us point out that in general

The union of two events A and B, denoted $A \cup B$, is the event which consists of all the outcomes (points) contained in A, in B, or in both.

Thus, if A is the event that a secretary will get a raise and B is the event that she will get a promotion, then $A \cup B$ is the event that she will get a raise or a promotion, including the possibility that she will get both. (Incidentally, $A \cup B$ reads "the union of A and B," "A union B,"

"*A cup B*," or simply "*A or B*," and in some books on probability it is denoted $A+B$.)

Similarly, we shall write the event that *U and S both occur* as $U \cap S$, and refer to it as the **intersection** of *U* and *S*. In general,

> **The intersection of two events A and B, denoted $A \cap B$, is the event which consists of all the outcomes (points) that A and B have in common.**

Thus, with reference to the preceeding example, $A \cap B$ is the event that the secretary will get a raise as well as a promotion. Also, if *C* is the event that a high school football team will win its first game and *D* is the event that it will win its second game, then $C \cap D$ is the event that it will win its first two games. (In this case, $A \cap B$ reads "*the intersection of A and B*," "*A intersection B*," "*A cap B*," or simply "*A and B*," and it is sometimes denoted $A \cdot B$ or AB.)

Finally, we shall write the event that *R does not occur* as R', and refer to it as the **complement** of *R*. In general,

> **The complement of an event A, denoted A', is the event which consists of all the outcomes (points) of the sample space that are not contained in A. In other words, A' is the event that A will not occur.**

Thus, with reference to the example which we used to illustrate the meaning of $A \cup B$ and $A \cap B$, A' is the event that the secretary will *not* get a raise and B' is the event that she will *not* get a promotion. (In practice, A' reads "the complement of *A*," "*A* prime," or "non-*A*," and in some books it is denoted $\sim A$, $-A$, or $\bar{A}$.)

One advantage of the set notation is that we can explain new ideas symbolically, without having to go through lengthy verbal details. For instance, to explain what we mean by "*A* and *B* are mutually exclusive," we can now say that this is the case when $A \cap B = \varnothing$, where $\varnothing$ is the empty set which does not contain any outcomes (points) at all. Thus, *A* and *B* are mutually exclusive when their intersection is empty, namely, when they have no outcomes (points) in common, and this is precisely how we defined "mutually exclusive" on page 92.

Although we shall use set notation in this chapter and the next, **Boolean Algebra** (namely, the **Algebra of Sets**) will not be used to any appreciable extent. This is because most relationships among sets, for instance, the relationship $(A \cup B)' = A' \cap B'$ which tells us that "neither *A* nor *B*" has the same meaning as "not *A* and not *B*," can be pictured by means of **Venn diagrams** like those of Figures 4.7 through 4.10. In these diagrams, named

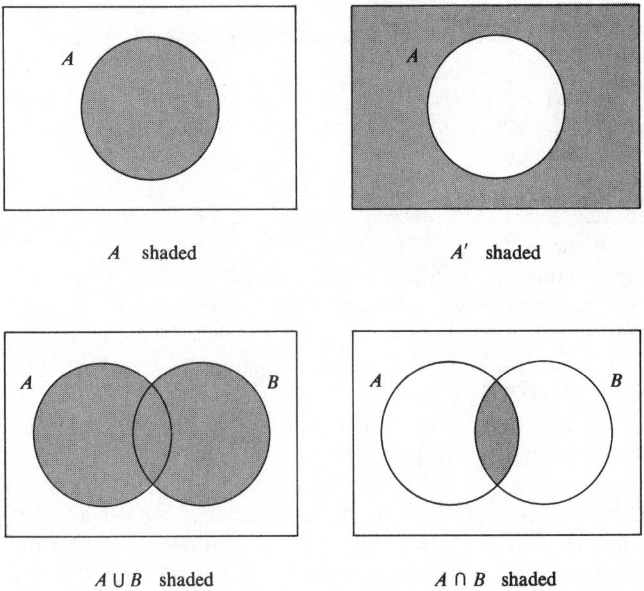

FIGURE 4.7 Venn diagrams.

after the British mathematician and logician John Venn (1834–1923), the sample space is represented by a rectangle, while subsets (or events) are represented by regions within the rectangles, usually circles or parts of circles. Thus, the shaded regions of the four diagrams of Figure 4.7 represent, respectively, event A, the complement of event A, the union of two events A and B, and the intersection of two events A and B. To give an application, if A is the event that a new movie is a financial success and B is the event that it is an artistic success, the regions shaded in the four Venn diagrams of Figure 4.7 represent, respectively, the event that the movie is a financial success, the event that the movie is not a financial success, the event that the movie is a financial success and/or an artistic success, and the event that the movie is a financial success as well as an artistic success.

To illustrate how Venn diagrams can be used to *verify* relationships among sets, and thus avoid the necessity of having to give formal proofs, let us refer to the relationship mentioned in the preceeding paragraph, namely, $(A \cup B)' = A' \cap B'$. In Figure 4.8, $A \cup B$ is represented by the region *inside* the heavy line, and hence its complement $(A \cup B)'$ is represented by the shaded region *outside* the heavy line. In Figure 4.9, A' is represented by the horizontally ruled region outside circle A, B' is represented

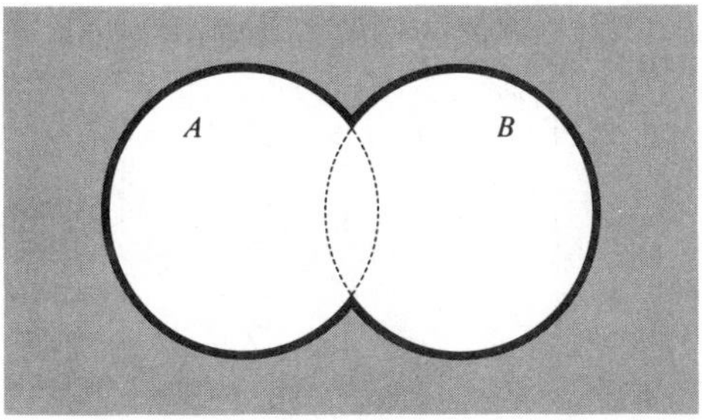

$(A \cup B)'$

FIGURE 4.8 Venn diagram.

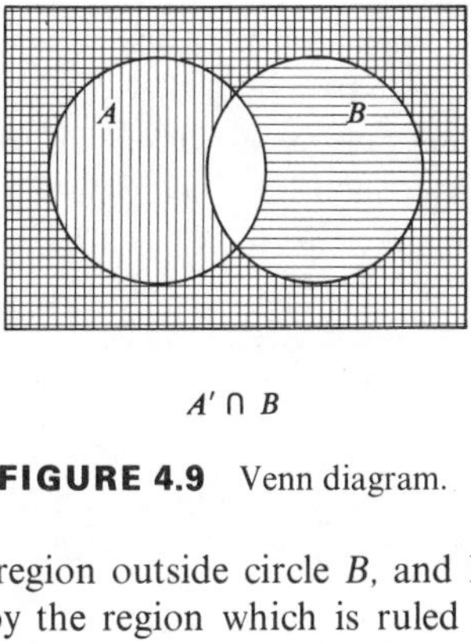

$A' \cap B$

FIGURE 4.9 Venn diagram.

by the vertically ruled region outside circle B, and hence the intersection $A' \cap B'$ is represented by the region which is ruled *both ways.* As can be seen by inspection, this region is the same as the one which we shaded in Figure 4.8, and this demonstrates, or verifies, that $(A \cup B)'$ is the same event as $A' \cap B'$. With reference to the illustration of the preceding paragraph, this tells us that "The movie is not a financial or artistic success" means the same as "The movie is not a financial success and it is not an artistic success," and as we indicated earlier, $(A \cup B)'$ can also be read as "The movie is neither a financial success nor an artistic success."

If we want to represent three events in one Venn diagram, we usually draw the circles as in Figure 4.10, where A, B, and C might be the events that the aforementioned movie is a financial success, that it is an artistic success, and that it is rated GP. As can be seen from Figure 4.10, the sample space is thus divided into eight regions (numbered 1 through 8),

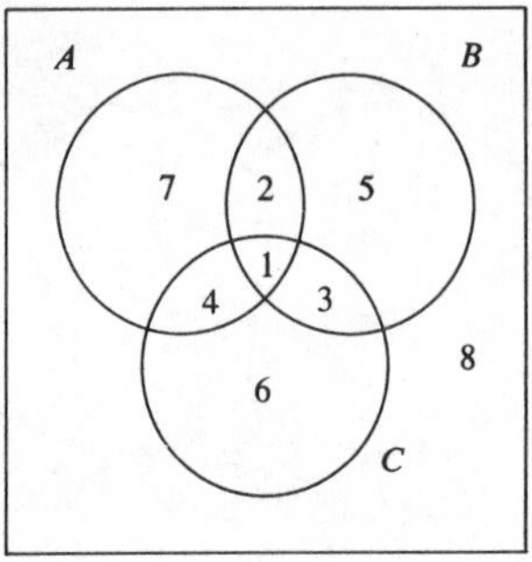

FIGURE 4.10 Venn diagram.

which are easy to identify with the corresponding events. Region 3, for example, is outside A, inside B, and inside C, so that it represents the event that the movie is not a financial success, but an artistic success and rated GP. Similarly, Region 7 is inside A, outside B, and outside C, so that it represents the event that the movie is a financial success, but neither an artistic success nor rated GP. It will be left to the reader to identify some of the other regions (and also some of their combinations) in Exercise 15 below.

EXERCISES

1. If W is the event that a saleslady working at a certain department store is satisfied with the working conditions and M is the event that she is satisfied with her wages, express *in words* each of the following events:
(a) W'; (c) $W \cup M$; (e) $M' \cup W'$;
(b) M'; (d) $W' \cap M'$; (f) $M \cap W'$.

2. If E is the event that a given candidate for Congress will be elected and K is the event that he will keep his promises, express *symbolically* each of the following events:
(a) The candidate will not be elected.
(b) The candidate will be elected and keep his promises.
(c) The candidate will not be elected or he will keep his promises.
(d) The candidate will be elected but not keep his promises.
(e) The candidate will be elected or he will not keep his promises.
(f) The candidate will neither be elected nor keep his promises.

3. A publishing firm plans to build a warehouse in Southern California, and its management has to decide between sites in Los Angeles, San

Diego, Ontario, Pasadena, Santa Barbara, Riverside, Santa Monica, and Burbank. If A represents the event that they will choose a site in San Diego or Santa Barbara, B represents the event that they will choose a site in San Diego or Ontario, C represents the event that they will choose a site in Santa Barbara or Riverside, and D represents the event that they will choose a site in Los Angeles or Santa Barbara, list the elements (site selections) of each of the following sets:

(a) A'; (d) $B \cap C$; (g) $C \cup D'$;
(b) D'; (e) $B \cup C$; (h) $(B \cup C)'$;
(c) $C \cap D$; (f) $A \cup B$; (i) $B' \cap C'$.

4. With reference to the sample space of Figure 4.5, list the points of the sample space which comprise the following events:

(a) B'; (c) $A \cup D$; (e) $C' \cap D'$;
(b) C'; (d) $A \cap B$; (f) $B' \cap C$.

5. With reference to the illustration on page 93 and Exercise 6 on page 95, list the points of the sample space which comprise the following events:

(a) S'; (d) $S \cap W$; (g) $R' \cap T$;
(b) W'; (e) $V \cup X$; (h) $S' \cup V'$;
(c) $R \cup U$; (f) $R \cap W$; (i) $T \cap X'$.

6. Express *in words* the events which are represented symbolically in the nine parts of Exercise 5.

7. If we let 1 and 0 represent *heads* and *tails,* the results obtained in three flips of a coin can be represented by means of points such as (1, 0, 1) and (0, 1, 1), where the first represents *heads, tails, heads,* and the second represents *tails, heads, heads.*

(a) List the eight points of the corresponding sample space and also present them geometrically in a diagram similar to that of Figure 4.3.
(b) If F is the event that the first two flips yield *heads,* G is the event that at least one of the flips is *tails,* H is the event that there are either three *heads* or three *tails,* and I is the event that in the second flip the coin comes up *heads,* list the points of the sample space belonging to each of the corresponding subsets.
(c) With reference to part (b), express *in words* the events G', I', $H \cup F$, $G \cap I$, $F \cap G'$, and $H' \cap I'$.
(d) List the points which comprise the six events of part (c).

8. With reference to Exercises 8 and 9 on pages 95 and 96, list the points of the sample space which constitute the following events:

(a) K'; (d) $M \cap P$; (g) $O \cap P'$;
(b) O'; (e) $K \cup N$; (h) $M \cup P'$;
(c) $L \cup N$; (f) $L \cap M$; (i) $O' \cap L'$.

9. Express *in words* the events which are represented symbolically in the nine parts of Exercise 8.

10. With reference to Exercises 11 and 12 on page 96, list the points of the sample space which comprise the following events:

(a) V';
(b) Q';
(c) $R \cup S$:
(d) $T \cap W$;
(e) $U \cap Q$;
(f) $T \cup W$;
(g) $R' \cup T'$;
(h) $Q' \cap W'$;
(i) $S' \cap U$.

11. Express *in words* the events which are expressed symbolically in the nine parts of Exercise 10.

12. Use Venn diagrams to verify that $(A \cap B)' = A' \cup B'$ for any two sets A and B. This relationship, as well as the one verified on page 101, are referred to as the two *de Morgan laws*. If A is the event that a person is wealthy and B is the event that he resides in New York, express $(A \cap B)'$ as well as $A' \cup B'$ in words.

13. Use Venn diagrams to verify the two *distributive laws*
(a) $A \cap (B \cup C) = (A \cap B) \cup (A \cap C)$;
(b) $A \cup (B \cap C) = (A \cup B) \cap (A \cup C)$.

14. Use Venn diagrams to verify that the following relationships hold for any two sets A and B:
(a) $A \cup (A \cap B) = A$;
(b) $A \cap (A \cup B) = A$;
(c) $(A \cap B) \cup (A \cap B') = A$;
(d) $A \cup B = (A \cap B) \cup (A' \cap B) \cup (A \cap B')$.

15. With reference to the example on page 102 which pertains to Figure 4.10, list the region or regions which represent the following events:
(a) The movie is a financial success and rated GP, but not an artistic success.
(b) The movie is a financial and artistic success.
(c) The movie is rated GP but is not an artistic success.
(d) The movie is a financial success and/or an artistic success.
(e) The movie is neither an artistic success nor rated GP.

16. With reference to the example on page 102 which pertains to Figure 4.10, explain *in words* what events are represented by the following regions of the Venn diagram:
(a) Region 1;
(b) Region 2;
(c) Region 6;
(d) Region 8;
(e) Regions 1 and 3 together;
(f) Regions 3 and 5 together;
(g) Regions 2, 5, and 7 together;
(h) Regions 1, 3, 4, and 6 together;
(i) Regions 3, 5, 6, and 8 together.

17. Suppose that a soap manufacturer wants to advertise on television and that F is the event that the program he chooses will appeal to teenagers, G is the event that it will be on network television, and H is the event that it will get a high rating. With reference to the Venn diagram of Figure 4.11, list the regions or combinations of regions which represent the following events:

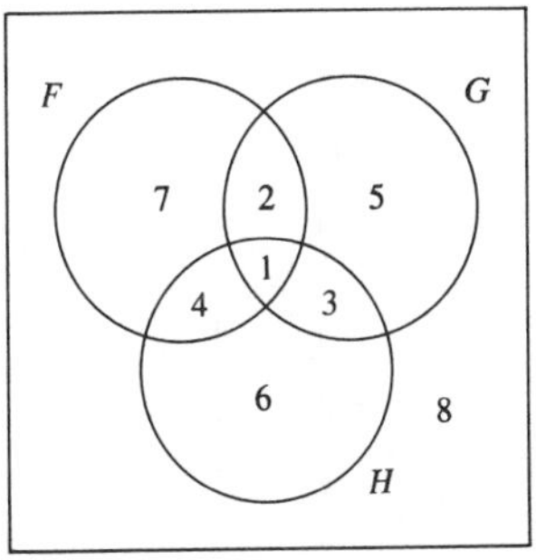

FIGURE 4.11 Venn diagram for Exercise 17.

(a) The soap manufacturer will choose a program with appeal to teenagers, which will neither be on network television nor get a high rating.
(b) The soap manufacturer will choose a program with appeal to teenagers, which will be on network television.
(c) The soap manufacturer will choose a program which will not appeal to teenagers, yet have a high rating.
(d) The soap manufacturer will choose a program which will not appeal to teenagers, and neither will it be on network television nor will it get a high rating.
(e) The soap manufacturer will choose a program which will get a high rating.

18. With reference to Exercise 17 and the Venn diagram of Figure 4.11, explain *in words* what events are represented by the following regions:

(a) Region 1;
(b) Region 4;
(c) Regions 1 and 3 together;
(d) Regions 2 and 7 together;
(e) Regions 3, 5, and 6 together;
(f) Regions 4, 6, 7, and 8 together;
(g) Regions 1, 2, 3, 4, 5, and 6 together;
(h) Regions 1, 3, 4, 5, 6, and 8 together.

19. With reference to Exercise 17 and the Venn diagram of Figure 4.11, list the regions or combinations of regions which represent the following events:

(a) F'; (c) $F \cap H$; (e) $F' \cap G'$;
(b) $F \cup G$; (d) $G \cup H'$; (f) $F \cap (G' \cap H)$.

20. Instead of Venn diagrams, we sometimes use **Euler diagrams** (named after the Swiss mathematician *Lenhard Euler* (1707–1788)), in which the circles can be drawn so that one is inside or outside the other. What special relationships between the events A and B are expressed by means of the two Euler diagrams of Figure 4.12? Also, if A is the

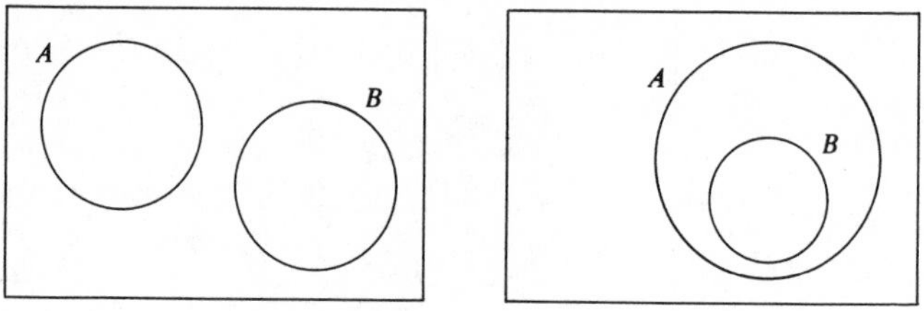

FIGURE 4.12 Euler diagrams.

event that a person is intelligent and B is the event that he is a prominent citizen, what (not necessarily true) relationships between intelligent persons and prominent citizens are expressed by (a) the first Euler diagram of Figure 4.12, and (b) the second Euler diagram of Figure 4.12. (Actually, we shall not use Euler diagrams in this book.)

THE NUMBER OF ELEMENTS IN A SET

As we have used the term, the *elements* of a set are the objects, or "things," which it contains, and in our discussion they have been possibilities, outcomes, or points. Since counting possibilities is important not only in the classical concept of probability but whenever we are dealing with random selections which may be regarded as equally likely for one reason or another, let us illustrate how Venn diagrams can be used in counting possibilities, namely, determining the number of elements in a set. Suppose, for instance, that one juror is to be selected by lot from a panel of 239 persons, of which 151 are women. Also, 77 of the 239 persons on the panel are under thirty years of age, and 48 of the 151 women are under thirty. *What we would like to know is the probability that the person selected will be a man who is at least thirty years old.*

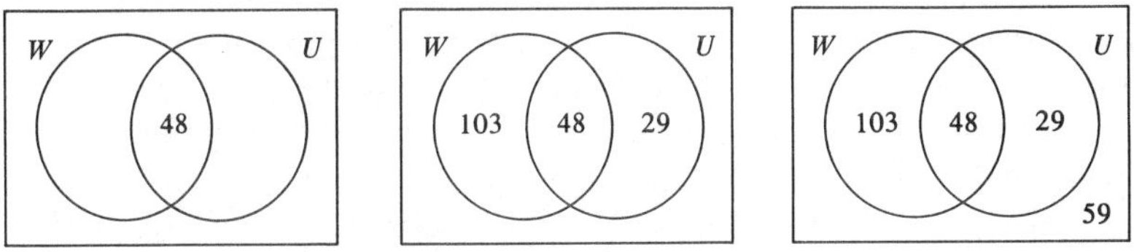

FIGURE 4.13 Venn diagrams.

Drawing circles to represent women (set W) and persons under thrity (set U) as in the Venn diagrams of Figure 4.13, we begin by writing 48 in the region common to the two circles of the first diagram. Then we observe that $151-48 = 103$ of the possibilities representing women must be in W but outside U and $77-48 = 29$ of the possibilities representing persons under thirty must be in U but outside W, and these figures are shown in the corresponding regions of the second Venn diagram. This accounts for $103+48+29 = 180$ of the possibilities, and it leaves $239-180 = 59$ for the region outside the two circles, as is indicated in the third of the Venn diagrams of Figure 4.13. Since the selection of the juror is by lot, we conclude that the probability is $\frac{59}{239}$, or almost 0.25, that it will be a man who is at least thirty years old.

To consider another example, suppose that the manager of a ski resort has the following information: 72 percent of his guests stay at least three days, 77 percent spend at least \$25 per day, 70 percent are satisfied with the food, 60 percent stay at least three days and spend at least \$25 per day, 59 percent stay at least three days and are satisfied with the food, 58 percent spend at least \$25 per day and are satisfied with the food, and 54 percent stay at least three days, spend at least \$25 per day, and are satisfied with the food. What we would like to know are the percentages that (a) *a guest will be satisfied with the food, spend at least \$25 per day, but stay less than three days;* (b) *a guest will stay at least three days, spend less than \$25 per day, and dislike the food;* and (c) *a guest will stay less than three days, spend less than \$25 per day, and dislike the food.*

This example differs from the preceding one in that we shall be working with percentages instead of numbers, and will need three circles instead of two. Drawing the circles representing guests who stay at least three days (set D), guests who spend at least \$25 per day (set M), and guests who are satisfied with the food (set F) as in the Venn diagram of Figure 4.14, we begin by writing 54 in the region common to D, M, and F. Then we observe that in $60-54=6$ percent of the possibilities a guest stays at least three days, spends at least \$25 per day, but is not satisfied with the food; in $59-54=5$ percent of the possibilities a guest stays at least three days, is satisfied with the food, but spends less than \$25 per day; and in $58-54=4$ percent of the possibilities a guest spends at least \$25 per day, is satisfied with the food, but stays for less than three days. Filling in these percentages in the appropriate regions of the Venn diagram of Figure 4.14, we find that we have found the answer to the first of the three questions asked on page 107—*the percentage of guests who are satisfied with the food, spend at least \$25 per day, but stay less than three days is 4.*

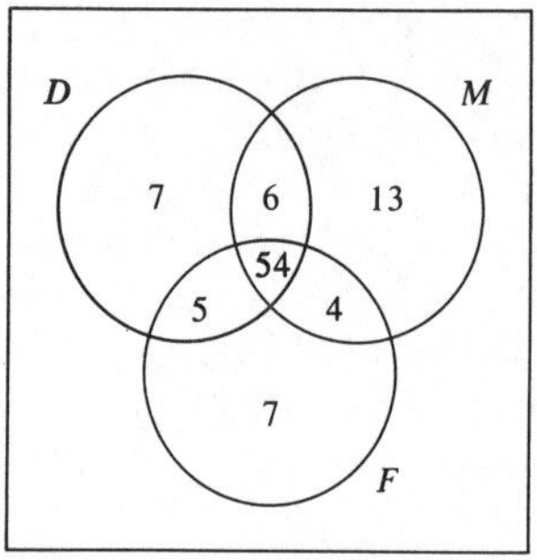

FIGURE 4.14 Venn diagram.

To continue, we observe that in $77-(6+54+4)=13$ percent of the possibilities a guest spends at least \$25 per day, but stays less than three days, and is dissatisfied with the food; in $70-(5+54+4)=7$ percent of the possibilities a guest is satisfied with the food, but stays less than three days, and spends less than \$25 per day; and in $72-(5+54+6)=7$ percent of the possibilities a guest stays at least three days, but spends less than \$25 per day, and is dissatisfied with the food. These figures are also shown in the Venn diagram of Figure 4.14, and we find that the answer to question (b) is 7, which is *the percentage of persons who stay at least three days, but spend less than* \$25 *per day, and dislike the food.* Finally, adding all the percentages shown in the Venn diagram of Figure 4.14, we get $7+6+13+5+54+4+7=96$, and we conclude that in the remaining 4 percent of the possibilities a guest stays less than three days, spends less than

$25 per day, and dislikes the food. Thus, the answer to question (c) is that *the desired percentage is 4.*

EXERCISES

1. A company has 240 employees, of which 145 got a raise, 85 got a promotion, and 50 got both. If one of the employees is to be chosen by lot to head a labor-management committee, what is the probability that he will have received neither a raise nor a promotion?

2. In a group of 200 college students, 77 are enrolled in a course in World History, 64 are enrolled in a course in World Geography, and 92 are *not* enrolled in either course. How many of these students must be enrolled in both courses?

3. A market research organization claims that among 400 housewives interviewed in a large city, 312 regularly look at the food ads in the daily paper, 248 regularly read the "Dear Abby" column, 173 regularly look at both, and 43 look at neither on a regular basis. Draw a Venn diagram and fill in the figures which correspond to the various regions to check whether the results of this survey should be questioned.

4. The manager of a baseball park knows from experience that 37 percent of all persons attending games buy a program, 54 percent buy peanuts, 67 percent buy a soft drink, 23 percent buy a program and peanuts, 34 percent buy peanuts and a soft drink, 28 percent buy a program and a soft drink, and 16 percent buy a program, peanuts, and a soft drink. What percentages of persons attending a baseball game at this park will
(a) buy a program and a soft drink, but no peanuts;
(b) buy peanuts, but neither a program nor a soft drink;
(c) buy either peanuts or a soft drink;
(d) buy a soft drink, but no peanuts;
(e) buy neither a program, nor peanuts, nor a soft drink?

5. Among 60 houses advertised for sale there are eight with swimming pools, three or more bedrooms, and wall-to-wall carpeting, five with swimming pools and three or more bedrooms, but no wall-to-wall carpeting, three with swimming pools and wall-to-wall carpeting, but fewer than three bedrooms, eight with swimming pools but neither wall-to-wall carpeting nor three or more bedrooms, 24 with three or more bedrooms but neither a swimming pool nor wall-to-wall carpeting, two with three or more bedrooms and wall-to-wall carpeting, but no swimming pool, three with wall-to-wall carpeting, but neither a swimming pool nor three or more bedrooms, and seven without any of these

features. If one of these houses is to be chosen at random to be featured in a television commercial, what is the probability that
(a) it will have wall-to-wall carpeting;
(b) it will have a swimming pool;
(c) it will have neither a swimming pool nor wall-to-wall carpeting?

5

RULES OF PROBABILITY

INTRODUCTION

Everybody feels that he "knows" what numbers are—one, two, three, four, and so on—but it is not an easy task even for a competent mathematician to give rigorous definitions. In fact, it was not until very recently that anyone even tried. The same is true also in geometry, where everybody has his intuitive ideas about such things as points and lines. Again, everybody feels that he "knows" what they are, yet how can they be defined? A point on a blackboard is only a deposit of small particles of chalk, and when we draw a line we are creating something which may look like a mountain range to a very small insect.

One thing we are taught in modern mathematics is that everything cannot be defined, for to define something we have to define it in terms of something else, and this leads to a chain which never ends. The only escape from this difficulty is to leave certain basic terms *undefined,* and this we do, for example, with the term "set," which has assumed such a popular role in modern mathematics. Of course, everybody "knows" what we mean by a set—a "collection" of objects or a "group" of things—yet we shall not embarrass the reader by asking what might be meant here by "collection" or "group."

Strictly speaking, it is true that many things in mathematics have to be left undefined, which once prompted a great English logician to comment with tongue in cheek that "Mathematicians never know what they are talking about." It should be understood, however, that even though we may leave certain things undefined, we do narrow things down considerably by specifying rules according to which these undefined "mathematical objects" must behave. For instance, in modern mathematics we are taught at the high school level, or even earlier, that *sets* are "mathematical objects" which behave according to the rules of *Boolean Algebra* (the *Algebra of Sets*), and even if the reader is not familiar with this kind of algebra, he must surely have heard about such things as the *commutative laws* or the *associative laws* which characterize the behavior of "ordinary" numbers. [The *commutative law for addition* tells us, for example, that $2+3 = 3+2$, and the *associative law for multiplication* tells us, similarly, that $2\cdot(3\cdot5) = (2\cdot3)\cdot5$.] The same is also true in plane geometry, where we may not be able to define points or lines, yet specify such rules as "There is one and only one line through any two distinct points" or "There is only one line through a given point which is parallel to a given line."

We have made these general observations because they apply also to the study of probability. Probabilities *are* difficult to define, and as we saw in Chapter 2, most everyone seems to have his own intuitive ideas. This suggests that it may be best to leave the term "probability" undefined, subject only to the restriction that probabilities are "mathematical objects" which must obey certain rules. These rules, basically very simple ones, will be introduced in the next section, and they lead to the mathematical *Theory of Probability,* or the *Calculus of Probability* as it is sometimes called. In contrast to Chapter 2, this part of the study of probability is *not controversial,* for as we shall see, the rules we shall introduce are compatible with the classical probability concept, the frequency interpretation, as well as the subjective approach. Indeed, the fact that probabilities "behave" in the same way mathematically regardless of how they may have been interpreted in Chapter 2, is what we hinted at on page 57.

THE POSTULATES OF PROBABILITY

The rules according to which "mathematical objects" must behave are generally referred to as **axioms** or **postulates**. Since the first of these terms is used mainly when such rules are looked upon as *self-evident truths,* while the second is used when the rules are more in the nature of *assumptions,* we shall speak here of the **Postulates of Probability**. There is nothing sacred about these postulates, but they lead to very useful mathematical concepts, namely, to concepts which are applicable to the world in which we live.

To formulate the three postulates of probability and the mathematical theory to which they lead, we shall continue the practice of denoting events by means of capital letters, and we shall write the probability of an event A as $P(A)$, the probability of an event B as $P(B)$, and so on. Also, we shall follow the common practice of denoting the set of all possibilities, namely, the sample space, with the letter S. As we shall formulate them here, the three postulates apply only when the sample space S is *finite,* but the modification which is required, for example, to handle the coin-flipping example on page 91, where the sample space is *infinite,* will be given on page 193. The first two postulates are:

> **POSTULATE 1: The probability of any event must be a positive real number or zero; symbolically, $P(A) \geqq 0$ for any event A.**

POSTULATE 2: The probability of a sample space is always 1; symbolically, $P(S) = 1$ for any sample space S.

As we have said, there is nothing sacred about any of these postulates—they merely reflect some of the properties which are desirable, we feel, when it comes to measuring uncertainties. Like measurements of height and weight, for example, we do not want such measurements to be negative, and this is the reason for Postulate 1. Note that this postulate is satisfied by all three of the probability concepts of Chapter 2: In the classical concept the ratio of the number of successes to the total number of possibilities is a real number (a fraction, in fact) which cannot be negative, and the same holds for the ratio $\frac{a}{a+b}$ used in the subjective approach to define probabilities. Finally, in the frequency interpretation a *proportion in the long run* has to be a real number on the interval from 0 to 1.

So far as the second postulate is concerned, it merely sets a *convenient* upper limit which we identify with *certainty*; after all, it is certain that one of the outcomes in S must occur, and it is to this certainty that we assign the probability 1. We could have let $P(S) = 100$ or $P(S) = 35$, for example, but by letting $P(S) = 1$ we are automatically accommodating the various interpretations of Chapter 2. As we indicated in that chapter, certainty was identified with a probability of 1 for each of the three interpretations.

The third postulate of probability is especially important, but it is not quite so "intuitively obvious" as the other two.

POSTULATE 3: If two events A and B are mutually exclusive, the probability that one or the other will occur equals the sum of their respective probabilities; symbolically,

$$P(A \cup B) = P(A) + P(B)$$

for any two mutually exclusive events A and B.*

So far as the classical probability concept is concerned, this rule is certainly

* It will be assumed throughout that the events to which our formulas pertain are subsets of one and the same sample space. What can happen when this is not the case will be discussed briefly in the introduction to Chapter 6. To give an example, suppose that for one restaurant the probability is 0.15 that a customer will order wine with his meal while for another restaurant the probability is 0.22 that a customer will order a glass of beer. Evidently, we cannot conclude anything from this about the probability that a customer will order *wine or beer* with his meal in either or both of these restaurants (except that for the first restaurant this probability must be *at least* 0.15, and for the second restaurant it must be *at least* 0.22).

satisfied. If there are n equally likely possibilities in a given situation of which r comprise event A and s others comprise event B, then the $r+s$ possibilities together comprise event $A \cup B$, and we have $P(A) = \frac{r}{n}$, $P(B) = \frac{s}{n}$, and $P(A \cup B) = \frac{r+s}{n} = \frac{r}{n} + \frac{s}{n} = P(A) + P(B)$. For instance, the probability of drawing a queen from an ordinary deck of 52 playing cards is $\frac{4}{52}$, the probability of drawing a red king is $\frac{2}{52}$, and the probability of drawing either a queen or a red king is $\frac{4}{52} + \frac{2}{52} = \frac{6}{52}$. Also, in connection with the example on page 40, the probability is $\frac{6 \cdot 2 \cdot 6!}{8!} = \frac{3}{14}$ that if the student randomly arranges his eight textbooks on a shelf, exactly one other book will be between his two business texts, and if we combine this with the result obtained on page 40, namely that the probability is $\frac{1}{4}$ that the business texts will be together, we find that the probability is $\frac{1}{4} + \frac{3}{14} = \frac{13}{28}$ (or almost a half) that the two business texts will be together or have one other book between them.

To show that the third postulate is satisfied by the frequency interpretation, we have only to observe that if one event will occur, say, 13 percent of the time, and another event will occur, say, 28 percent of the time, and the two events cannot both occur at the same time (that is, they are mutually exclusive), then one or the other will occur $13 + 28 = 41$ percent of the time. Of course, the same argument holds also when we are dealing with proportions. Thus, if somebody is planning to buy a small imported car and the probabilities that he will buy a German import or a Japanese import are, respectively, 0.34 and 0.18, the probability that he will buy one or the other is $0.34 + 0.18 = 0.52$. Also, if a doctor's answering service knows that on a Saturday afternoon the probabilities are, respectively, 0.44 and 0.17 that the doctor will be playing golf at his club or cards at the house of his brother-in-law, they can figure that the probability is $0.44 + 0.17 = 0.61$ that he can be reached at either place. In each of these examples, either alternative can happen, but not both.

So far as subjective probabilities are concerned, it is important to note that it does *not* follow from our definition in Chapter 2 that they must satisfy Postulate 3. For instance, if a student feels that the odds against his getting an A in a course in American History are 4 to 1 and the odds against his getting a B are 3 to 1, it does *not* follow that he must also feel that the odds against his getting an A or a B are 11 to 9. This is what they would have to be to satisfy Postulate 3, for the probabilities that he *will* get an A or a B are, respectively, $\frac{1}{5}$ and $\frac{1}{4}$, and $\frac{1}{5} + \frac{1}{4} = \frac{9}{20}$. Thus, proponents of the subjective point of view impose the third postulate as a **consistency criterion**. This means that if a person's subjective probabilities "behave" in accordance with Postulate 3, he is said to be *consistent*; otherwise, he is said to be *inconsistent* and his probability judgment will have to be taken with a grain of salt. Of course, if a person makes only *one* subjective probability statement about a given situation it is impossible to judge whether or not he is consistent—after all, it takes the probabilities relating to two mutually exclusive events A and B to check whether $P(A \cup B)$ equals $P(A)+P(B)$. This is a serious handicap, for in most practical situations we do not have enough information to check for consistency.

It is easier to justify Postulate 3 when subjective probabilities are measured as in Chapter 3. To illustrate, suppose that the same student is asked how much he would take *now* in lieu of the uncertainty of a gift of \$20, which he is to receive only if he gets an A. If he feels that \$4 would be fair, we can argue that his expectation of $\$20 \cdot p$, where p is the probability that he will get an A, equals \$4, namely, that $20p = 4$, and hence that $p = \frac{4}{20} = \frac{1}{5}$. Similarly, if the student feels that \$5 would be a fair amount to settle for in lieu of a gift of \$20 which he is to receive only if he gets a B, it stands to reason that he should also be willing to settle for \$4+\$5 = \$9 in lieu of a gift of \$20 which he is to receive if he gets either an A or a B. This, of course, leads to the "correct" odds of 11 to 9, and it should be noted that when subjective probabilities are measured in this way, "reasonable behavior" means the same as "satisfying the consistency criterion." Unfortunately, subjective probabilities cannot always be measured in this way.

Postulate 3 applies only to two mutually exclusive events, but it can easily be generalized so that it applies also to more than two mutually exclusive events, namely, to more than two events of which only one can occur.

GENERALIZATION OF POSTULATE 3: If the k events $A_1, A_2, \ldots,$ and A_k are mutually exclusive, the probability that one of them will occur equals the sum of their respective probabilities; symbolically

$$P(A_1 \cup A_2 \cup \ldots \cup A_k) = P(A_1)+P(A_2)+ \ldots +P(A_k)$$

where $\cup$ may again be read as "or." With this formula we can now calculate the probability that one of a set of mutually exclusive events will occur. For instance, if the probabilities that a given high school graduate will enroll at U.C.L.A., San Diego State, or Stanford are, respectively, 0.09, 0.14, and 0.05, then the probability that he will enroll at one of these schools is $0.09+0.14+0.05 = 0.28$. Also, if Mr. Jones is planning to buy a new color television set and the probabilities are 0.16, 0.24, 0.11, and 0.18 that he will buy a Zenith, R.C.A., Motorola, or Magnavox, then the probabilities that he will buy one of these kinds of sets is $0.16+0.24+0.11+0.18 = 0.69$.

PROBABILITY MEASURES

When we assign probabilities to the various things that can happen in a given situation, we are said to be providing a **probability measure** for the corresponding sample space. In connection with this, it is important to note that the postulates of probability do *not* tell us how probabilities should be assigned to the various events—they merely restrict the ways in which it can be done, or to put it differently, *they tell us what cannot be done.* For instance, the following are two different ways in which probabilities *can be assigned* to the four mutually exclusive outcomes E, F, G, and H, of which one must happen in a given situation:

$$\text{(a)}\ P(E) = 0.11,\ P(F) = 0.27,\ P(G) = 0.52,\ P(H) = 0.10$$

$$\text{(b)}\ P(E) = \tfrac{3}{29},\ P(F) = \tfrac{12}{29},\ P(G) = \tfrac{8}{29},\ P(H) = \tfrac{6}{29}$$

None of the postulates are violated in these two examples, but the first postulate would be violated if we let

$$\text{(c)}\ P(E) = 0.25,\ P(F) = 0.38,\ P(G) = 0.46,\ P(H) = -0.09$$

and the second postulate would be violated if we let

$$\text{(d)}\ P(E) = 0.32,\ P(F) = 0.25,\ P(G) = 0.19,\ P(H) = 0.38$$

Note that $P(H)$ is negative in (c), and that $P(S) = P(E \cup F \cup G \cup H) = 0.32 + 0.25 + 0.19 + 0.38 = 1.14$ in (d). Of course, in actual practice probabilities are assigned on the basis of past experience, a careful analysis of all relevant factors, or assumptions, and the postulates merely provide a means of checking whether we are "playing the game according to the rules."

The job of assigning probabilities to *all* the events that are possible in a given situation can be very tedious, to say the least. For a sample space with as few as *five* points representing, say, the alternatives that a U.S.D.A. inspector will grade a carcass of beef "Prime," "Choice," "Good," "Commercial," or "Utility," there are already $2^5 = 32$ possible events in accordance with the rule of Exercise 24 on page 32. There is the event that he will grade it "Prime," the event that he will grade it "Good," the event that he will grade it "Prime or Choice," the event that he will grade it "Commercial or Utility," the event that he will grade it "Choice, Good, or Commercial," and so on. Things get worse very rapidly when a sample space has more points than in this example—for 10 points, for instance, there are $2^{10} = 1{,}024$ different subsets or events, and for 24 points there are

$2^{24} = 16{,}777{,}216$ possible events [as the reader was asked to show in part (b) of Exercise 24 on page 32].

Fortunately, it is seldom, if ever, necessary to assign probabilities to all possible events, for the following rule makes it possible to determine the probability of any event given only the probabilities which are assigned to the individual points of the sample space:

The probability of any event A is given by the sum of the probabilities of the individual outcomes comprising A.

A probability measure can thus be established without assigning a probability to each subset or event, by assigning probabilities only to the individual outcomes.

This very useful rule is illustrated by means of Figure 5.1, where the dots represent the individual (mutually exclusive) outcomes; the fact that the probability of A is given by the sum of the probabilities of the points comprising A follows immediately from the generalization of Postulate 3 which we gave on page 115.

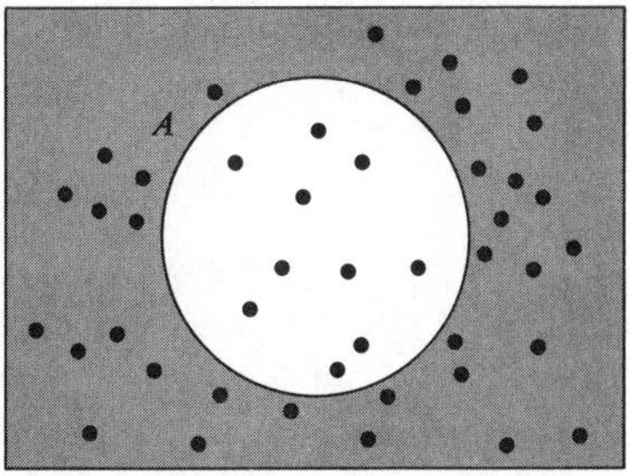

FIGURE 5.1 Venn diagram of a sample space.

To illustrate this rule, suppose that the probabilities that the U.S.D.A. inspector mentioned on page 116 will grade the beef "Prime," "Choice," "Good," "Commercial," or "Utility" are, respectively, 0.12, 0.37, 0.40, 0.07, and 0.04. Then, the probability that he will grade it "Prime or Choice" is $0.12+0.37 = 0.49$, the probability that he will grade it "Choice, Good, or Commercial" is $0.37+0.40+0.07 = 0.84$, and the probability to any other event can be determined in the same way. To give another example, let us refer back to page 88 where two students were asked about a new record, and let us suppose that the nine points of the sample space (shown

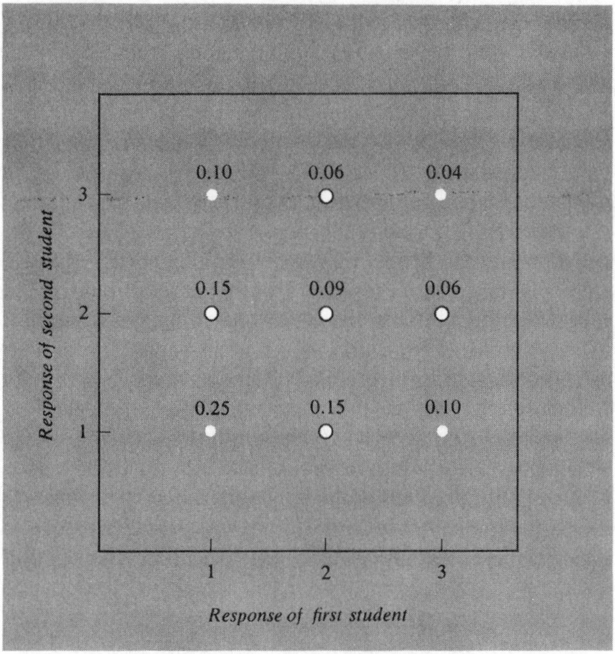

FIGURE 5.2 Sample space for responses of two students.

originally in Figure 4.2) have the probabilities assigned to them in Figure 5.2. Altogether, this sample space has $2^9 = 512$ different subsets or events, and according to the above rule we can find the probabilities of any of them by simply adding the probabilities assigned to the respective points which they contain. For instance, if we want to know the probability that at least one of the two students will dislike the new record, we have only to add the probabilities assigned to the five points circled in Figure 5.2, and we get $0.15+0.06+0.09+0.15+0.06 = 0.51$. Similarly, if we want to know the probability of event D on page 91, that one of the students will like the new record and the other won't care, we have only to add the probabilities assigned to the corresponding two points, and we get $0.10+0.10 = 0.20$. Continuing in the same way, the reader will be asked to verify in Exercise 18 on page 122 that the probabilities of events A, B, and C on page 91 are, respectively, 0.50, 0.30, and 0.64.

The situation becomes even simpler when the individual outcomes are all *equiprobable*, that is, when each of the n points of a sample space is assigned the probability $\frac{1}{n}$. In that case, the sum of the probabilities assigned to the s points which comprise a given event A is

$$\underbrace{\frac{1}{n}+\frac{1}{n}+\ldots+\frac{1}{n}}_{s \text{ terms}}=\frac{s}{n}$$

This, of course, is nothing but the formula which we introduced on page 37 in connection with the classical approach, but lest there be misunderstandings, it should be clear that the result which we have obtained here does *not necessarily* constitute support for the classical concept of probability. All we have shown is that the classical approach is *compatible* with the postulates of probability, which is nothing new, and when it comes to justifying the equiprobability of outcomes, we are as much in the dark as before.

EXERCISES

1. Analyzing business conditions in general, four government officials make the following claims: The first claims that the probabilities for unemployment to go up, remain unchanged, or go down are, respectively, 0.51, 0.33, and 0.12; the second claims that the respective probabilities are 0.55, 0.49, and −0.04; the third claims that the respective probabilities are 0.52, 0.38, and 0.10; and the fourth claims that the respective probabilities are 0.48, 0.34, and 0.21. Comment on these claims.
2. An experiment has five possible (mutually exclusive) outcomes, Q, R, T, U, and V. Check whether the following assignments of probability are in accordance with the rules:
   (a) $P(Q)=\frac{1}{6}$, $P(R)=\frac{1}{6}$, $P(T)=\frac{1}{6}$, $P(U)=\frac{1}{6}$, $P(V)=\frac{1}{6}$;
   (b) $P(Q)=\frac{1}{3}$, $P(R)=\frac{1}{24}$, $P(T)=\frac{1}{6}$, $P(U)=\frac{1}{4}$, $P(V)=\frac{5}{24}$;
   (c) $P(Q)=0.16$, $P(R)=0.18$, $P(T)=0.20$, $P(U)=0.22$, $P(V)=0.24$;
   (d) $P(Q)=0.13$, $P(R)=0.34$, $P(T)=-0.02$, $P(U)=0.25$, $P(V)=0.30$;
   (e) $P(Q)=0.26$, $P(R)=0.22$, $P(T)=0.24$, $P(U)=0.21$, $P(V)=0.18$.
3. A football coach claims that the odds are 3 to 2 that his team will win an upcoming game, while the odds against his team losing or tieing the game are, respectively, 4 to 1 and 9 to 1. Do the corresponding probabilities satisfy the three postulates of probability?
4. The manager of a gas station has found that 58 percent of his customers ask him to fill their tank, while 22 percent ask him to

check under the hood. From this he concludes that the odds are 4 to 1 that a customer will ask him to fill his tank or check under the hood. Discuss the validity of this argument.

5. Asked about his chances of promotion, a college instructor replies that the odds are 2 to 1 that he will not be an assistant professor five years from now, and the odds are 5 to 1 that he will not be an associate professor five years from now. Furthermore, he feels that it is an even bet (the odds are 1 to 1) that he will be either an assistant professor or an associate professor five years from now. Discuss the consistency of the corresponding probabilities.

6. An author feels that the odds against his new novel becoming a great success are 7 to 1 and the odds against it becoming a modest success are 3 to 1. To be consistent, what should he consider fair odds against his novel becoming either a great success or a modest success?

7. There are two Ferraris in a race, and an expert feels that the odds against their winning are, respectively, 2 to 1 and 3 to 1. Furthermore, he claims that there is a better-than-even chance that either of the two Ferraris will win. Discuss the consistency of these predictions.

8. At the end of a golf tournament three players are tied for the lead so that there will have to be a play-off, to see who will get the first prize of $20,000, who will get the second prize of $12,000, and who will get the third prize of $8,000. If one of the players feels that the odds are 4 to 1 against his taking first place in the play-off and that he would, in fact, settle for $12,400 right now, what probability is he thus assigning to his taking the second prize of $12,000? (*Hint*: Letting p denote the probability that he will win the second prize, express the probability that he will take the third prize in terms of p, equate his mathematical expectation to $12,400, and solve for p.)

9. Asked about the chances that the weather will improve or remain the same, a weatherman predicts that the odds are 2 to 1 that the weather will improve and 5 to 1 that it will remain the same. What is wrong with these odds?

10. Discuss the following statement: If the odds are 1 to 4 that a couple will have only one child and 2 to 5 that they will have two children, then the odds are $1+2=3$ to $4+5=9$ that they will have either one child or two children.

11. Sometimes a college student has her lunch at the school cafeteria, sometimes she brings her own lunch, sometimes she has her lunch at a nearby restaurant, sometimes she goes home for lunch, and sometimes she skips lunch altogether to lose weight. If the corresponding probabilities are 0.23, 0.39, 0.15, 0.16, and 0.07, find

(a) the probability that she will have lunch at the cafeteria, bring her own lunch, or have lunch at the restaurant;
(b) the probability that she will bring her own lunch, go home for lunch, or skip lunch to lose weight;
(c) the probability that she will not have lunch at either the cafeteria or the restaurant.

12. If a certain psychology professor asks a question of one of his students (chosen at random from a very large class), the probabilities that this student will have received an A, B, C, D, or F in the last examination are, respectively, 0.09, 0.23, 0.36, 0.18, and 0.14. What are the probabilities that the student received
(a) at least a C;
(b) at least a D;
(c) a B, a C, or a D;
(d) at most a C?
Assume that each student in the class has taken the examination.

13. The probability that a review board will rate a given movie X, R, GP, or G are, respectively, 0.01, 0.24, 0.36, and 0.39. Find the probabilities that they will rate the movie
(a) X or R;
(b) GP or G;
(c) R, GP, or G.

14. Mr. Clark needs new tires for his car and the probabilities are 0.17, 0.22, 0.08, 0.29, 0.21, and 0.03, respectively, that he will buy Uniroyal tires, Goodyear tires, Michelin tires, General tires, Goodrich tires, or Seiberling tires. Find the probabilities that he will buy
(a) Goodyear or Goodrich tires;
(b) Uniroyal, General, or Goodyear tires;
(c) Michelin or Seiberling tires;
(d) Goodyear, General, Goodrich, or Seiberling tires.

15. A waiter knows from experience that the probabilities are 0.15, 0.22, 0.09, 0.10, 0.04, and 0.08, respectively, that a customer will order chocolate cake, apple pie, cherry pie, ice cream, sherbet, or watermelon for dessert. What are the probabilities that a customer will order
(a) apple pie or cherry pie;
(b) ice cream or sherbet;
(c) chocolate cake, ice cream, or watermelon;
(d) cherry pie, ice cream, or sherbet;
(e) chocolate cake, apple pie, cherry pie, or watermelon;
(f) none of these desserts?
Assume that these desserts cannot be ordered in combinations.

16. According to a teacher's experience, the probabilities that a student will misspell zero, one, two, three, or *four or more* words in a short essay are, respectively, 0.11, 0.28, 0.25, 0.20, and 0.16. Find the probabilities that a student will misspell
(a) at least three words;
(b) at most one word;
(c) anywhere from one through three words;
(d) fewer than four words.

17. In a certain city, the probabilities that a driver will receive zero, one, two, three, four, or *five or more* traffic citations within one year are, respectively, 0.27, 0.39, 0.23, 0.06, 0.04, and 0.01. Find the probabilities that a driver will receive
(a) one or two traffic citations;
(b) at most one traffic citation;
(c) anywhere from two through four traffic citations;
(d) at least two traffic citations.

18. Verify the values given on page 118 for the probabilities of events A, B, and C of Figure 4·5 on page 92.

19. Referring to the sample space of Figure 4.6 and assuming that each point has the same probability of $\frac{1}{15}$, find
(a) the probabilities of events R, S, T, and U as defined on page 93;
(b) the probabilities of events V, W, and X as defined in Exercise 6 on page 95;
(c) the probabilities of the events which are given in the nine parts of Exercise 5 on page 103.

20. Suppose that in Exercise 8 on page 95 the points (0, 0), (1, 0), (1, 1), (2, 0), (2, 1), (2, 2), (3, 0), (3, 1), (3, 2), (3, 3), (4, 0), (4, 1), (4, 2), (4, 3), and (4, 4) of the sample space are assigned the probabilities 0.01, 0.04, 0.04, 0.09, 0.16, 0.09, 0.05, 0.15, 0.16, 0.08, 0.01, 0.03, 0.05, 0.03, and 0.01. Find
(a) the probabilities of events K, L, and M as defined in that exercise;
(b) the probabilities of events N, O, and P as defined in Exercise 9 on page 96;
(c) the probabilities of the events which are given in the nine parts of Exercise 8 on page 103.

21. Suppose that in Exercise 11 on page 96 each of the 20 points of the sample space has the same probability of $\frac{1}{20}$. Find
(a) the probabilities of events Q, R, and S as defined in that exercise;
(b) the probabilities of events T, U, V, and W as defined in Exercise 12 on page 96;
(c) the probabilities of the events which are given in the nine parts of Exercise 10 on page 104.

22. Suppose that in Exercise 14 on page 96 each of the single applicants has the probability $\frac{1}{10}$ of getting the job, and each of the married applicants has the probability $\frac{1}{5}$. Find the probabilities of the events described in the four parts of that exercise.

23. Suppose that in Exercise 15 on page 97 the probabilities are 0·15, 0.12, 0.08, 0.21, 0.17, 0.10, 0.04, and 0.13, respectively, that the dealer will first sell Car 1, Car 2, . . ., or Car 8. Find the probabilities that the first car he sells will
(a) have no airconditioning;
(b) have power steering;
(c) be at least two years old.

24. Suppose that in Figure 4.10 the events represented by Regions 1, 2, 3, and 5 are assigned probabilities of $\frac{1}{12}$, while the events represented by Regions 4, 6, 7, and 8 are assigned probabilities of $\frac{1}{6}$. Find the probabilities of the five events described in parts (e) through (i) of Exercise 16 on page 104.

25. Suppose that in Figure 4.11 the events represented by Regions 1 and 2 are assigned probabilities of 0.06, the events represented by Regions 3 and 5 are assigned probabilities of 0.14, the events represented by Regions 4 and 7 are assigned probabilities of 0.09, and the events represented by Regions 6 and 8 are assigned probabilities of 0.21. Find
(a) the probabilities of events F, G, and H as defined in Exercise 17 on page 105;
(b) the probabilities of the events described in parts (a) through (e) of Exercise 17 on page 105;
(c) the probabilities of the events given in parts (a) through (f) of Exercise 19 on page 105.

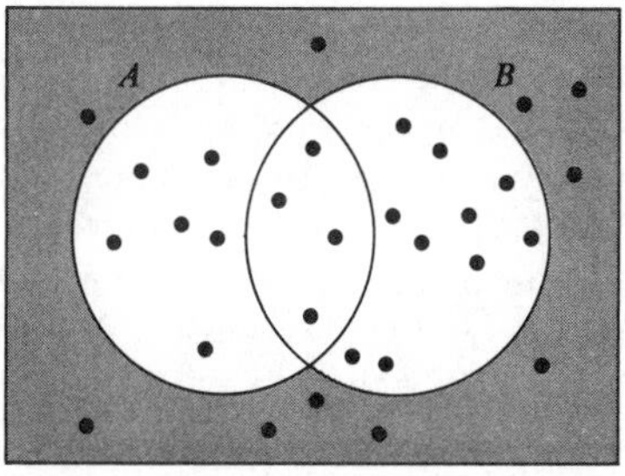

FIGURE 5.3 Diagram for Exercise 26.

26. If each of the points of Figure 5.3 represents an outcome having the probability $\frac{1}{30}$, find
(a) $P(A)$; (c) $P(A \cup B)$; (e) $P(A')$; (g) $P(A' \cap B)$;
(b) $P(B)$; (d) $P(A \cap B)$; (f) $P(A \cap B')$; (h) $P(A' \cup B')$.

SOME FURTHER RULES

Starting with the three postulates, we can obtain many further rules about the mathematical "behavior" of probabilities. Some of these are "intuitively obvious," some are very easy to prove, some are very complicated, yet virtually all of them have important applications. For instance, among the immediate consequences of the three postulates we find that *the probability that an event will occur and the probability that it will not occur always add up to* 1, *probabilities can never be greater than* 1, and *events that cannot possibly occur have the probability zero.* Symbolically,

$$P(A) + P(A') = 1 \quad \textit{for any event } A$$

$$P(A) \leqq 1 \quad \textit{for any event } A$$

and

$$P(\varnothing) = 0$$

where A' is the *complement* of A and $\varnothing$ is the *empty set,* which we mentioned earlier on page 91. So far as the first rule is concerned, we already showed in Chapter 2 that it holds regardless of whether we look at probabilities from the classical point of view, in accordance with the frequency interpretation, or subjectively. Of course, that does not constitute a proof, but we can take care of this with the following argument: Since A and A' between them include all the points of the sample space, we can write $A \cup A' = S$, where S denotes the sample space with which we are concerned, and hence $P(A \cup A') = P(S)$. Then, $P(S) = 1$ according to Postulate 2, and since A and A' are by definition mutually exclusive, $P(A \cup A') = P(A) + P(A')$ according to Postulate 3. Thus, $P(A \cup A') = P(S)$ leads to $P(A) + P(A') = P(S) = 1$, and this completes the argument.

The rule we have just proved is sometimes written in the form $P(A') = 1 - P(A)$, so that $P(A')$ is expressed directly in terms of $P(A)$. For instance, if the probability that we will be late for work is 0.26, the probability that we will *not* be late is $1 - 0.26 = 0.74$; if the probability of rolling 7 or 11 with a pair of dice is $\frac{2}{9}$, the probability of *not* rolling 7 or 11 is $1 - \frac{2}{9} = \frac{7}{9}$; and if the probability that a given golf professional will win the U.S. Open is 0.12, the probability that he will *not* win this tournament is $1-0.12=0.88$.

The following is a principle which the reader may well recall from other courses in mathematics:

> **Once a rule has been proved, it can be used as an argument in the proof of other rules.**

This simply means that we do not always have to go back to the basic postulates, and we shall thus use the rule $P(A')=1-P(A)$ to prove the second of the rules given above, namely, the rule that $P(A) \leqq 1$ for any event A. The kind of argument we shall use is called *reductio ad absurdum*—we assume the opposite of what we hope to prove, show that it leads to a contradiction, and hence conclude that the opposite of what we assumed must be true. Since we want to show that $P(A) \leqq 1$ for any event A, let us assume that there exists some event A for which $P(A)$ is greater than 1. Then, since $P(A')=1-P(A)$, the assumption that $P(A)$ exceeds 1 implies that $P(A')$ is *negative* (for we are subtracting from 1 a number greater than 1), and since this cannot be by virtue of Postulate 1, it follows that our assumption must have been wrong. In other words, there can be no event whose probability exceeds 1. So far as the various probability concepts of Chapter 2 are concerned, this simply expresses the fact that the number of favorable outcomes cannot exceed the total number of possibilities, an event cannot occur more than 100 percent of the time, and the ratio $\frac{a}{a+b}$, where a and b are positive amounts bet for and against an event, cannot exceed 1.

The third rule on page 124, which the reader will be asked to prove in Exercise 1 on page 129, simply expresses the fact that the probability 0 is assigned to any event which cannot possibly occur. Clearly, if there are no favorable outcomes the probability of a success is $\frac{0}{n}=0$; in the frequency interpretation an event which cannot occur happens 0 percent of the time; and so far as subjective probabilities are concerned, we already discussed this matter on page 53. Actually, in practice we also assign 0 probabilities to events which are so unlikely that, colloquially speaking they "would not happen in a million years." For instance, we would assign the probability 0 to the event that a monkey set loose on a typewriter might by chance type Plato's *Republic* word for word. Similarly, we would not hesitate to assign a probability of 0 to the event that a golfer will shoot three "holes in one" in a row, or to the event that it will snow in Phoenix, Arizona, in the month of July. Logically speaking, these things *can* happen, but they are so unlikely that they may be treated as if they were impossible.

GENERAL ADDITION RULES

The third postulate of probability and its generalization on page 115 are sometimes referred to as **special addition rules**; they are "special" in that they apply only to mutually exclusive events. They cannot be used, for example, to find the probability that an American tourist will visit Paris or Rome on a trip to Europe, the probability that a person will break an arm or a leg in an automobile accident, or the probability that a housewife will buy steak, lamb chops, or chicken on her next trip to the supermarket. In the first case, the tourist can visit Paris as well as Rome, in the second case the person can break an arm as well as a leg, and in the third case the housewife can buy two of the items, or all three.

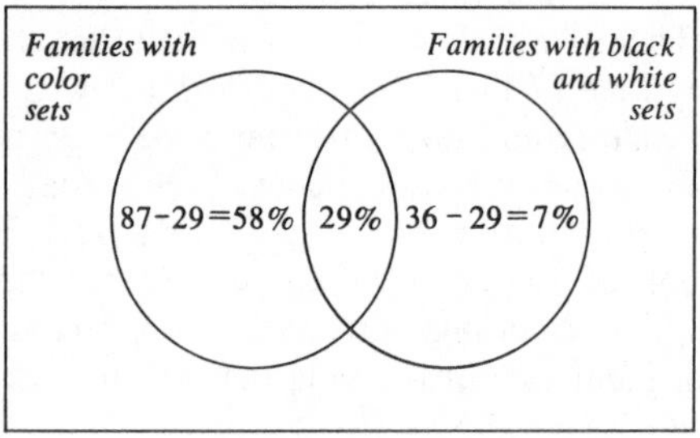

FIGURE 5.4 Venn diagram.

To obtain an addition rule which holds also for events that are not mutually exclusive, let us take a brief look at a survey conducted in a certain metropolitan area which showed that 87 percent of all families own a color television set and 36 percent own a black and white television set. Can we conclude from this that $87+36 = 123$ percent of the families have either kind of set? Of course not, and this is not only because the figure exceeds 100, but *the two percentages should never have been added because owning a color set and owning a black and white set are not mutually exclusive events.* To find the correct answer, we would also have to know what percentage of the families own both kinds of sets—at least one of each kind. If the answer is 29 percent, we can proceed as is illustrated in Figure 5.4: $87-29 = 58$ percent of the families own a color set but no black and white set, $36-29 = 7$ percent of the families own a black and white set but no color set, and hence $58+29+7 = 94$ percent of the families own either kind of set or both.

This result could also have been obtained by *subtracting* from the original total of $87+36=123$ the 29 percent which was erroneously included *twice*—once in the percentage of families with color sets and once in the percentage of families with black and white sets. In fact, if we translate all these percentages into proportions and let C and B denote the respective events that a randomly chosen family in the given metropolitan area will own a color set or a black and white set, we can express the preceding argument symbolically by writing

$$\begin{aligned} P(C \cup B) &= P(C)+P(B)-P(C \cap B) \\ &= 0.87+0.36-0.29 \\ &= 0.94 \end{aligned}$$

More generally, let us now show that $P(A \cup B) = P(A)+P(B)-P(A \cap B)$ for any two events A and B. Assigning, as in Figure 5.5, the probabilities a, b, c, and d to the four regions into which the two circles divide the

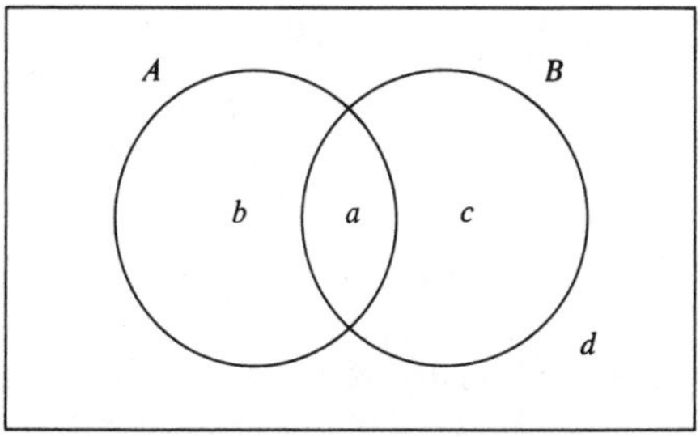

FIGURE 5.5 Venn diagram.

Venn diagram, we find by inspection that $P(A) = b+a$, $P(B) = a+c$, $P(A \cap B) = a$, and $P(A \cup B) = b+a+c$, and hence that

$$\begin{aligned} P(A)+P(B)-P(A \cap B) &= (b+a)+(a+c)-a \\ &= b+a+c \\ &= P(A \cup B) \end{aligned}$$

This demonstrates what we set out to show, namely, that

GENERAL ADDITION RULE: For any two events A and B,

$$P(A \cup B) = P(A)+P(B)-P(A \cap B)$$

Note that when A and B *are* mutually exclusive, $P(A \cap B) = 0$ since two mutually exclusive events cannot both occur, and the new formula reduces to that of Postulate 3.

To give another illustration of the general addition rule, let us refer again to the two-student-interview example and the probabilities of Figure 5.2. If events A and B are defined as on page 91, namely, as the events that "the first student likes the new record" and "the second student dislikes it," we find that $P(A) = 0.10+0.15+0.25 = 0.50$, $P(B) = 0.15+0.09+0.06 = 0.30$, $P(A \cap B) = 0.15$, and hence

$$P(A \cup B) = 0.50+0.30-0.15 = 0.65$$

This is the probability that the first student interviewed will like the new record and/or the second student will dislike it, and it can be checked by adding directly the probabilities of the corresponding five points in Figure 5.2.

It is not too difficult to extend the general addition rule so that it applies also to three or more events. For three events A, B, and C, the formula becomes

$$\begin{aligned} P(A \cup B \cup C) = {} & [P(A)+P(B)+P(C)] \\ & -[P(A \cap B)+P(A \cap C)+P(B \cap C)] \\ & +P(A \cap B \cap C) \end{aligned}$$

and it should be observed that first we take the sum of the probabilities of the individual events, then we *subtract* the sum of the probabilities of all pairwise intersections, and finally we *add* the probability of their three-way intersection $A \cap B \cap C$. The proof of this formula, which in principle is very similar to the proof which we gave on page 127, will be left to the reader in Exercise 16 on page 132.

To illustrate the use of the formula for $P(A \cup B \cup C)$, suppose that if a person visits his dentist, the probability that he will have his teeth cleaned is 0.46, the probability that he will have a cavity filled is 0.22, the probability that he will have a tooth extracted is 0.19, the probability that he will have his teeth cleaned and a cavity filled is 0.09, the probability that he will have his teeth cleaned and a tooth extracted is 0.11, the probability that he will have a cavity filled and a tooth extracted is 0.06, and the probability that he will have his teeth cleaned, a cavity filled, and a tooth extracted is 0.03. Substituting all these values into the formula of the preceding paragraph, we find that the probability that he will have his teeth cleaned and/or a cavity filled, and/or a tooth extracted is

$$(0.46+0.22+0.19)-(0.09+0.11+0.06)+0.03 = 0.64$$

To continue this example, the reader will be asked to show in Exercise 18 on page 132 that the probability is 0.44 that a person visiting his dentist will have *one and only one* of these things done to him.

EXERCISES*

1. Making use of the fact that $S \cup \varnothing = S$ for any sample space S, and S and $\varnothing$ are mutually exclusive (by default), prove the last of the three rules on page 124, namely that $P(\varnothing) = 0$.

2. Making use of the fact that $(A \cap B) \cup (A \cap B') = A$, which the reader was asked to verify by means of a Venn diagram in part (c) of Exercise 14 on page 104, show that $P(A) \geqq P(A \cap B)$ for any two events A and B.

3. Making use of parts (c) and (d) of Exercise 14 on page 104, show that $P(A \cup B) \geqq P(B)$ and also $P(A \cup B) \geqq P(A)$ for any two events A and B.

4. If set A is contained in set B (that is, if A is a subset of B), then the region of a corresponding Venn diagram which is *inside* the circle representing A and *outside* the circle representing B must be empty; symbolically, $A \cap B' = \varnothing$. Making use of this result as well as part (c) of Exercise 14 on page 104, show that $P(A) \leqq P(B)$ when A is contained in B.

5. Explain why there must be a mistake in each of the following statements:

(a) On Thursday, Harry has to take final examinations in psychology and economics. The probability that he will pass the psychology examination is 0.38, the probability that he will pass both examinations is 0.23, and the probability that he will pass the psychology examination but fail the economics examination is 0.16.

(b) The probability that the Internal Revenue Service will audit Mr. Jones' tax return is 0.17, and the probability that it will audit either Mr. Jones' tax return or that of his brother is 0.14.

(c) In discussing an upcoming swimming meet, a sports writer says that the odds are 3 to 1 that a native Californian will win the 100 meter free-style race, and 4 to 1 that the winner will be a native of San Francisco.

* The first four exercises, which are of a theoretical nature, may be omitted without loss of continuity.

(d) The probability that a football team will win its first game is 0.63, the probability that it will win its second game is 0.84, and the probability that it will win both games is 0.45. (*Hint*: Draw a Venn diagram and fill in the probabilities associated with the various regions.)

(e) Mrs. Miller has two daughters. The probability that her older daughter will get married within a year is 0.27, and the probability that both her daughters will get married within a year is 0.32.

(f) The probability that a person with a fever will get a shot from his doctor is 0.48, the probability that he will get some medication but no shot is 0.36, and the probability that he will get no medication and no shot is 0.12.

6. Given two mutually exclusive events A and B for which $P(A) = 0.27$ and $P(B) = 0.46$, find

(a) $P(A')$; (e) $P(A \cap B')$;
(b) $P(B')$; (f) $P(A' \cap B)$;
(c) $P(A \cup B)$; (g) $P(A' \cap B')$;
(d) $P(A \cap B)$; (h) $P(A' \cup B')$.

(*Hint*: Draw a Venn diagram and fill in the probabilities associated with the various regions.)

7. Given two events C and D for which $P(C) = 0.67$, $P(D) = 0.23$, and $P(C \cap D) = 0.12$, find

(a) $P(C')$; (d) $P(C \cap D')$;
(b) $P(D')$; (e) $P(C \cup D)$;
(c) $P(C' \cap D)$; (f) $P(C' \cap D')$.

(*Hint*: Draw a Venn diagram and fill in the probabilities associated with the various regions.)

8. The probability that a person stopping at a given gas station will ask to have his battery checked is 0.23, the probability that he will have the tires checked is 0.18, and the probability that he will have his battery and his tires checked is 0.08. What are the probabilities that a person stopping at this gas station will

(a) have his battery but not his tires checked;
(b) have his battery and/or his tires checked;
(c) have neither his battery nor his tires checked;
(d) have his battery or tires, but not both, checked?

9. To raise money for a worthy cause, a club sells raffle tickets numbered from 1 to 200. What are the probabilities that the number drawn will be

(a) divisible by 4;
(b) divisible by 5;
(c) divisible by 20?

Also use the results of parts (a), (b), and (c) to find the probability that the number drawn will be

(d) divisible by either 4 or 5.

10. For married couples living in the suburb of a large city, the probability that the husband will vote in a local election is 0.18, the probability that his wife will vote in the election is 0.23, and the probability that they will both vote is 0.14. What is the probability that at least one of them will vote?

11. A businessman has two secretaries. The probability that the one he hired most recently will be absent on any given day is 0.08, the probability that the other secretary will be absent on any given day is 0.07, and the probability that they will both be absent on any given day is 0.02.

(a) What is the probability that either or both secretaries will be absent on any given day?

(b) What is the probability that at least one of the two secretaries will show up for work on any given day?

(c) What is the probability that only one of the two secretaries will show up for work on any given day?

12. A teacher feels that the odds are 3 to 2 against her getting a promotion, it is an even bet that she will get a raise, and the odds are 4 to 1 against her getting both. What are the odds that she will get a promotion and/or a raise?

13. Among the 45 members of a coin club who attend a meeting, 37 collect U.S. coins, 19 collect foreign coins, and 13 collect U.S. and foreign coins. If the door prize of a $5 gold piece is awarded by lot to one of the members present, what are the probabilities that it will be won by

(a) a member who collects U.S. coins;

(b) a member who collects foreign coins;

(c) a member who collects U.S. and foreign coins;

(d) a member who collects U.S. and/or foreign coins;

(e) a member who collects neither U.S. nor foreign coins (but, perhaps, medals or paper money)?

14. Referring to Exercise 8 on page 95, suppose that X is the event that either two or three cabs are out on a call, and Y is the event that three of the cabs are in operative condition. Using the probability measure given to the corresponding sample space in Exercise 20 on page 122, find $P(X)$, $P(Y)$, and $P(X \cap Y)$. Then use these results to determine $P(X \cup Y)$, and verify the result by directly adding the probabilities assigned to the corresponding seven points of the sample space.

15. Referring to Exercise 14 on page 96 and Exercise 22 on page 123, find the probabilities that the job is given to a native of Arizona, that the job is given to someone with an M.A. degree, and that the job is given to a native of Arizona with an M.A. degree. Then use these results to determine the probability that the job is given to someone who is a native of Arizona and/or holds an M.A. degree, and verify the result by subtracting from 1 the probability assigned to the only applicant who is not a native of Arizona and does not hold an M.A. degree.

16. In Figure 5.6 we have assigned the probabilities a, b, c, d, e, f, g, and h to the events represented by the eight regions into which the three circles divide the Venn diagram.

(a) Express $P(A)$, $P(B)$, $P(C)$, $P(A \cap B)$, $P(A \cap C)$, $P(B \cap C)$, and $P(A \cap B \cap C)$ in terms of a, b, c, d, e, f, g, and h.

(b) Use the results of part (a) to verify the formula for $P(A \cup B \cup C)$ on page 128.

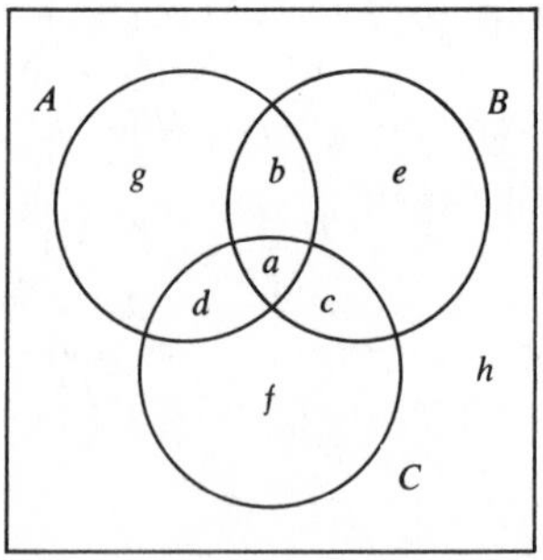

FIGURE 5.6 Venn diagram for Exercise 16.

17. Referring to Figure 4.11 and the probabilities assigned to the various regions in Exercise 25 on page 123, it can easily be seen that the event $F \cup G \cup H$ is represented by Regions 1 through 7 and, hence, that its probability is 0.79. Verify this result by determining $P(F)$, $P(G)$, $P(H)$, $P(F \cap G)$, $P(F \cap H)$, $P(G \cap H)$, and $P(F \cap G \cap H)$, and substituting these values into the formula for $P(F \cup G \cup H)$ on page 128.

18. Verify the value given on page 129 for the probability that a person visiting his dentist will have one and only one of the following things done to him: have his teeth cleaned, have a cavity filled, or have a tooth extracted.

19. Suppose that if a child visits Disneyland, the probability that he will go on the Jungle Cruise is 0.62, the probability that he will ride the Monorail is 0.69, the probability that he will go on the Matterhorn ride is 0.49, the probability that he will go on the Jungle Cruise and ride the Monorail is 0.37, the probability that he will go on the Jungle Cruise as well as the Matterhorn ride is 0.39, the probability that he will ride the Monorail and go on the Matterhorn ride is 0.34, and the probability that he will go on all three of these rides is 0.28. What is the probability that a person visiting Disneyland will go on at least one of these three rides?

20. Generalizing the description which we gave on page 128 of the formula for $P(A \cup B \cup C)$, let us state without proof that for any four events A, B, C, and D, the probability $P(A \cup B \cup C \cup D)$ can be obtained by taking the sum of the probabilities of the individual events, *subtracting* the sum of the probabilities of all pairwise intersections of the four events, *adding* the sum of the probabilities of all three-way intersections of the four events, and finally *subtracting* the probability of their four-way intersection $A \cap B \cap C \cap D$. Use this rule in the following problem: Suppose that if a person spends his vacation in a certain resort, the probability that he will play golf is 0.70, the probability that he will go swimming is 0.64, the probability that he will play tennis is 0.58, the probability that he will go horseback riding is 0.58, the probability that he will play golf and go swimming is 0.45, the probability that he will play golf and tennis is 0.42, the probability that he will play golf and go horseback riding is 0.41, the probability that he will go swimming and play tennis is 0.35, the probability that he will go swimming and horseback riding is 0.39, the probability that he will play tennis and go horseback riding is 0.32, the probability that he will play golf and tennis and go swimming is 0.23, the probability that he will play golf and go swimming and horseback riding is 0.26, the probability that he will play golf and tennis and go horseback riding is 0.21, the probability that he will play tennis and go swimming and horseback riding is 0.20, and the probability that he will take part in all four of these activities is 0.12. What is the probability that a person visiting this resort will take part in at least one of these four activities?

6

CONDITIONAL PROBABILITIES

INTRODUCTION

Consider the following argument: The probability of getting *heads* in the flip of a balanced coin is $\frac{1}{2}$, the probability of rolling a 6 with a balanced die is $\frac{1}{6}$, and the probability of getting heads with a balanced coin or a 6 with a balanced die is, therefore, $\frac{1}{2}+\frac{1}{6}=\frac{2}{3}$. This would be nice, if it were only true; in fact, we could then add the probability of drawing a *spade* from an ordinary deck of 52 playing cards, which is $\frac{1}{4}$, and come up with the comfortable odds of 11 to 1 that we will get either heads, a 6, or a spade. The fallacy of this argument is that when we determined the probabilities of getting heads with a balanced coin, a 6 with a balanced die, or a spade, we were referring to *different sample spaces,* and when this is the case, Postulate 3 and its generalizations cannot be applied. To ask for the probability of getting "heads, a 6, or a spade," we would have to invent a new game, perhaps by first determining by lot whether

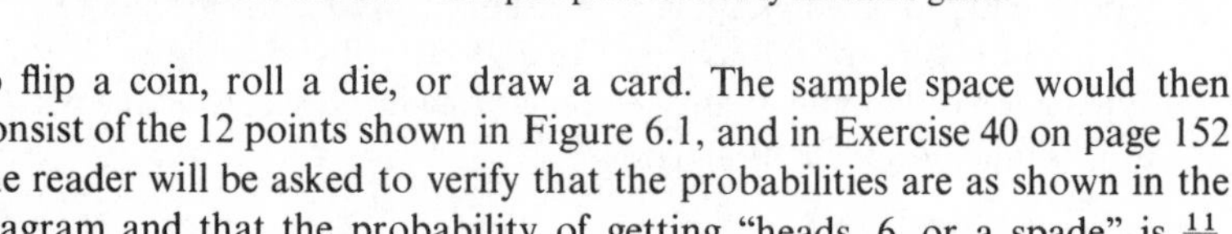

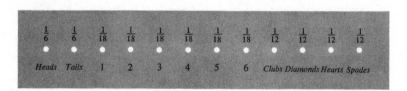

FIGURE 6.1 Sample space for newly invented game.

to flip a coin, roll a die, or draw a card. The sample space would then consist of the 12 points shown in Figure 6.1, and in Exercise 40 on page 152 the reader will be asked to verify that the probabilities are as shown in the diagram and that the probability of getting "heads, 6, or a spade" is $\frac{11}{36}$. In case the reader feels a bit uneasy about this example, let us point out that it has been our purpose to demonstrate that *the choice of the sample space which is appropriate in a given situation can be quite difficult and misleading.*

Difficulties can also arise when sample spaces are not adequately defined.

Without clarification it would be meaningless (or at least very confusing) to ask, say, for the probability that 10 years after graduating from college a teacher will make \$12,000 a year. *There are many different answers, and they can all be correct.* One of them might apply to all certified elementary school or high school teachers in the United States, another might include college teachers, another might refer to all teachers in North and South America, another might refer only to teachers employed in public schools, and so on. Since the choice of the sample space (namely, the set of all possibilities under consideration) is rarely self-evident, it is helpful to use the symbol $P(A|S)$ to denote the **conditional probability** of the event A relative to a given sample space S, or as we often call it "the probability of A *given* S." This idea is really nothing new, but it makes it explicitly clear that we are referring to a particular sample space S, and this is preferable to the notation we have used so far, unless the tacit choice of S is clearly understood. The notation $P(A|S)$ is of special value when we want to refer to more than one sample space in one and the same example. For instance, if A is the *event* that we get exactly two heads, T is the *sample space for three flips of a coin,* and F is the *sample space for four flips of a coin,* then $P(A|T)$ and $P(A|F)$ are both probabilities of getting exactly two heads, but the notation makes it clear that the first pertains to three flips of a coin while the second pertains to four flips of a coin.

CONDITIONAL PROBABILITY

To elaborate on the idea of a conditional probability, let us consider the following problem. Suppose that a newspaper editor has received 120 letters from irate readers about the controversial firing of a high school teacher, but has the space to print only one of these letters, presumably chosen at random. Suppose, furthermore, that the actual breakdown into letters written by students and parents, and letters supporting the teacher or the superintendent (who fired the teacher) is as shown in the following table:

	Written by students	*Written by parents*
Supporting the teacher	16	44
Suporting the superintendent	8	52
	24	96

More conveniently, perhaps this situation can be pictured by means of the Venn diagram of Figure 6.2, where C denotes the event that the letter which the editor will print came from a student, and T denotes the event that it will support the teacher. As can be seen from the Venn diagram or the table, the chances are pretty slim that the editor will print a letter written by a student in support of the teacher—symbolically, we can write $P(C \cap T) = \frac{16}{120} = \frac{2}{15}$.

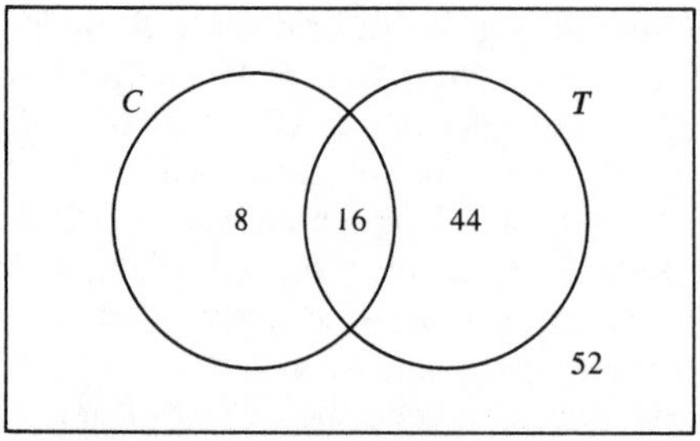

FIGURE 6.2 Venn diagram of sample space.

It is easy to see how this can lead to misunderstandings, as it fails to reveal the fact that the students who did write to the editor supported the teacher by a ratio of 16 to 8 (or 2 to 1). To bring this out, we need the conditional probability $P(T \mid C)$, namely, the probability that the letter which the editor selects will favor the teacher *given that it was written by a student*. This probability pertains to the *reduced* sample space C, pictured in Figure 6.3, which consists of only part of the original sample space shown in Figure 6.2. If we assume that the 24 possibilities in C are still equally likely, we get

$$P(T \mid C) = \frac{\text{number of possibilities in } C \cap T}{\text{number of possibilities in } C} = \frac{16}{24} = \frac{2}{3}$$

and this is more indicative, of course, of student sentiment about the teacher. Note that if we divide the numerator and denominator by 120 (the number of possibilities in the original sample space), we get

$$P(T \mid C) = \frac{16/120}{24/120} = \frac{P(C \cap T)}{P(C)}$$

and the conditional probability $P(T \mid C)$ has thus been expressed as the *ratio* of the probability that *the editor will print a letter written by a student in support of the teacher* to the probability that *he will print a letter which was written by a student.* This relationship is very important, and it should be observed that by looking at the sample space of Figure 6.3 instead of that of Figure 6.2, and hence at $\frac{P(C \cap T)}{P(C)}$ instead of $P(C \cap T)$, we have eliminated the distorting factor that so many more letters were written by parents than by students.

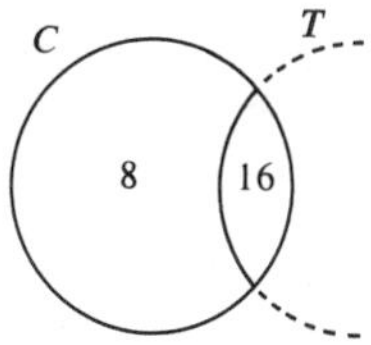

FIGURE 6.3 Venn diagram of reduced sample space.

Generalizing from the above example, let us now make the following definition of a **conditional probability:**

If A and B are events belonging to a given sample space S and $P(B) \neq 0$, then the conditional probability of event A relative to event B is given by

$$P(A \mid B) = \frac{P(A \cap B)}{P(B)}$$

Note that although we referred to equiprobable possibilities in our numerical example, this definition of conditional probabilities applies regardless of what probabilities are assigned to the individual outcomes. The only restriction is that $P(B)$ should not be zero, for division by zero is not allowed.

To consider an example where we are not dealing with equiprobable outcomes, suppose that the person in charge of a salvage operation feels that the probability tnat they will be able to locate the wreck of a sunken ship is 0.40, and that the probability that they will not only be able to locate it but raise it once it has been found is 0.36. What we would like to know is the probability that they will be able to raise the wreck *given that it has been found.* If we let F denote the event that they will be able to find the wreck and R the event that they will be able to raise it, the given information can be expressed as $P(F) = 0.40$ and $P(R \cap F) = 0.36$, and substitution into the formula for $P(R \mid F)$ yields

$$P(R \mid F) = \frac{P(R \cap F)}{P(F)} = \frac{0.36}{0.40} = 0.90$$

Thus, once the wreck has been found, the odds are 9 to 1 that it can be raised.

To introduce another concept which is important in the study of probability, let us refer again to the two-student-interview example, for which we showed on page 128 that $P(A) = 0.50$, $P(B) = 0.30$, and $P(A \cap B) = 0.15$, where A is the event that "the first student will like the new record" and B is the event that "the second student will dislike it." If we substitute the last two values into the formula for $P(A \mid B)$, which is now the conditional probability that the first student will like the new record given that the second student dislikes it, we get

$$P(A \mid B) = \frac{P(A \cap B)}{P(B)} = \frac{0.15}{0.30} = 0.50$$

and what is *special* about this result is that

$$P(A \mid B) = 0.50 = P(A)$$

This shows that the probability of event A is the same regardless of whether event B has occurred (occurs, or will occur), and we say that the event A is **independent** of event B. Informally, this means that the occurrence of A is is no way affected by the occurrence of B, and this should really not come as a surprise—there should be no relationship, or dependence, between the responses of two students chosen at random from a given group. It is of interest to note that in the first example of this section event T was *not* independent of event C. We showed that $P(T \mid C) = \frac{2}{3}$ and it can easily be seen from Figure 6.2 or the table on page 135 that $P(T) = \frac{16+44}{120} = \frac{1}{2}$. This is indicative of the fact that the editor's probability of choosing a letter favoring the teacher *increases* if he considers only letters sent in by the students.

When we said in the preceding paragraph that *the occurrence of A was in no way affected by the occurrence of B*, we could have added that *the occurrence of A would also not have been affected by the non-occurrence of B, nor would the occurrence of B have been affected by the occurrence or non-occurrence of A*. Indeed, it can be shown that whenever A is independent of B, then A is also independent of B', B is also independent of A, and B is also independent of A', so that when one of these is known to be true, we simply say that **A and B are independent**. To verify this for our example, for which we already showed on page 138 that $P(A \mid B) = P(A)$, we have only to refer to the Venn diagram of Figure 6.4, according to which

$$P(A \mid B') = \frac{P(A \cap B')}{P(B')} = \frac{0.35}{0.70} = 0.50 = P(A)$$

$$P(B \mid A) = \frac{P(B \cap A)}{P(A)} = \frac{0.15}{0.50} = 0.30 = P(B)$$

and

$$P(B \mid A') = \frac{P(B \cap A')}{P(A')} = \frac{0.15}{0.50} = 0.30 = P(B)$$

Let us add that if two events A and B are *not* independent, we say that they are **dependent**.

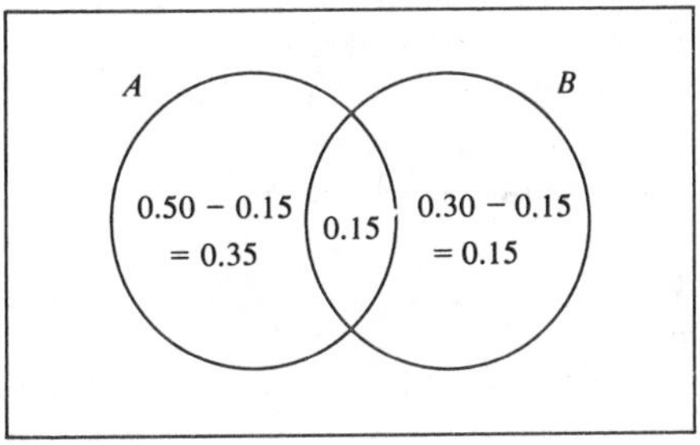

FIGURE 6.4 Venn diagram of sample space.

MULTIPLICATION RULES

An immediate consequence of our definition of a conditional probability is the following rule, called the **General Rule of Multiplication:**

For any two events A and B,

$$P(A \cap B) = P(B) \cdot P(A \mid B)$$

provided $P(B) \neq 0$.

In words, this rule states that *the probability that two events A and B will both occur is the product of the probability that one of them will occur and the probability that the other event will occur given that the first has occurred (occurs, or will occur).* Note that we did not specifically refer to A and B, and if we interchange these letters, the rule can also be written as

$$P(A \cap B) = P(A) \cdot P(B \mid A)$$

Actually, it takes only one step to prove the rule in its original form: We take the formula $P(A \mid B) = \frac{P(A \cap B)}{\mathrm{P(B)}}$, which defines the conditional probability $P(A \mid B)$, and we multiply the expressions on both sides of the equation by $P(B)$. To prove the rule in the second form, we simply interchange A and B and make use of the fact that $A \cap B = B \cap A$.

To illustrate this multiplication rule, let us consider the following situation: Being in a hurry to catch a plane, someone randomly grabs two paperback mysteries off a rack containing 15 paperback mysteries, four of which he had already read. What we would like to know is the probability that *he will have read neither of the mysteries which he grabs.* For the first book which he grabs we can argue that there are 11 successes and 15 possibilities, so that the probability of his getting a book which he has not yet read is $\frac{11}{15}$. This leaves four books he has read and 10 books he has not read, and it follows that the probability is $\frac{10}{14}$ that the second book which he grabs is also one he has not yet read. Thus, the probability that he has read neither book is

$$\frac{11}{15} \cdot \frac{10}{14} = \frac{11}{21}$$

and it may have occurred to the reader that this result can also be obtained by arguing that he can pick two of the 15 books in $\binom{15}{2} = 105$ ways, two of the books he has not yet read in $\binom{11}{2} = 55$ ways, and hence that the desired probability is

$$\frac{\binom{11}{2}}{\binom{15}{2}} = \frac{55}{105} = \frac{11}{21}$$

Using the multiplication rule, it can also be shown that the probability of his already having read both of the mysteries which he grabs is

$$\frac{4}{15}\cdot\frac{3}{14} = \frac{2}{35}$$

and it follows *by subtraction* that the probability of his having read only one of the two mysteries is

$$1 - \frac{11}{21} - \frac{2}{35} = \frac{44}{105}$$

To verify this last result by the methods of Chapters 1 and 2, we can argue that he can pick one of the books he has read in $\binom{4}{1}$ ways, one of the books he has not read in $\binom{11}{1}$ ways, and, hence, one of each kind in $\binom{4}{1}\cdot\binom{11}{1} = 4\cdot 11 = 44$ ways. As we have already shown that he can pick two of the 15 books in $\binom{15}{2} = 105$ ways, it follows that the probability of his having read only one of the two mysteries is

$$\frac{\binom{4}{1}\cdot\binom{11}{1}}{\binom{15}{2}} = \frac{44}{105}$$

and this agrees with the result which we obtained before.

To give an example where we are not dealing with equally likely possibilities, suppose that the odds are 3 to 1 that a rare tropical disease is diagnosed correctly, and once it is diagnosed correctly the probability is 0.80 that it will be cured. What we would like to know is the probability that *a person having this rare disease will have it diagnosed correctly and cured.* Clearly, if B is the event that the disease is diagnosed correctly and A is the event that it is cured, we have $P(B) = \frac{1}{1+3} = 0.25$, $P(A \mid B) = 0.80$, and substitution into the first form of the multiplication rule yields $P(A \cap B) = (0.25)(0.80) = 0.20$, which is the answer to our question.

When A and B are independent events, we can substitute $P(A)$ for $P(A \mid B)$ in the multiplication rule $P(A \cap B) = P(B)\cdot P(A \mid B)$, or $P(B)$ for $P(B \mid A)$ in the multiplication rule $P(A \cap B) = P(A)\cdot P(B \mid A)$, and we get the following rule, called the **Special Multiplication Rule** for two independent events:

$$P(A \cap B) = P(A)\cdot P(B)$$

This rule tells us that for independent events we simply multiply their respective probabilities. It can be used, for example, to find the probability of getting two heads in a row with a balanced coin or the probability of drawing two kings in a row from an ordinary deck of 52 playing cards (*provided* the first card is replaced before the second is drawn). For the two flips of the coin we get $\frac{1}{2}\cdot\frac{1}{2}=\frac{1}{4}$, which agrees with the observation that among the four possibilities—*heads* and *heads, heads* and *tails, tails* and *heads, tails* and *tails*—only one is a success. Similarly, for the two kings we get

$$\frac{4}{52}\cdot\frac{4}{52}=\frac{1}{169}$$

since there are four kings among the 52 cards. Note that if the first card is *not* replaced before the second is drawn, the probability that the second card will be a king *given that the first card was a king* is only $\frac{3}{51}$, and the multiplication rule on page 139 yields

$$\frac{4}{52}\cdot\frac{3}{51}=\frac{1}{221}$$

for the probability of drawing two kings. The distinction we have made here is very important in statistics, where we speak correspondingly of **sampling with replacement** and **sampling without replacement**. This will be discussed further in Chapter 7.

Now we are also in the position to answer a question raised in the beginning of Chapter 1. On page 4 we asked for the probability that the American League team will win a (best of seven) World Series after it is behind 3 games to 2. Since the only way in which the American League team can win the series is by winning the sixth and seventh games, the probability of its winning is $\frac{1}{2}\cdot\frac{1}{2}=\frac{1}{4}$ provided the two teams are evenly matched. Thus, the odds favoring the National League team are 3 to 1, even though we showed on page 4 that there are *two* possibilities favoring the National League team against *one* possibility favoring the American League team.

The special multiplication rule can easily be extended so that it applies also to the occurrence of three or more independent events—we simply *multiply all their respective probabilities.* For instance, the probability of getting three heads in a row with a balanced coin is $\frac{1}{2}\cdot\frac{1}{2}\cdot\frac{1}{2}=\frac{1}{8}$, the probability of drawing *with replacement* four spades in a row from an ordinary deck of 52 playing cards is $\frac{1}{4}\cdot\frac{1}{4}\cdot\frac{1}{4}\cdot\frac{1}{4}=\frac{1}{256}$, and the probability of getting a 3 in the first and fourth of five rolls of a balanced die, but not in the second, third, and fifth, is $\frac{1}{6}\cdot\frac{5}{6}\cdot\frac{5}{6}\cdot\frac{1}{6}\cdot\frac{5}{6}=\frac{125}{7{,}776}$, since the probabilities of rolling a 3 and not rolling a 3 are, respectively, $\frac{1}{6}$ and $\frac{5}{6}$.

In all these examples we assumed that the respective events were *independent*, namely, that *the occurrence of any one event or combination of events would not affect the occurrence of any other event.* This was perfectly reasonable, but strange things can happen when it comes to the independence of more than two events, as is illustrated in Exercise 37 on page 151, where an event A is independent of two events B and C taken individually, but not when they are taken together. This is why it is actually preferable to proceed the other way around, namely, *define* the independence of k events by requiring that the special multiplication rule apply to any 2, 3, ..., or k of the events.

For dependent events, the multiplication rules become somewhat more complicated, but stated in words they amount to the fact that we must take the *product* of the probability that *the first event will occur*, the probability that *the second event will occur given that the first event has occurred* (*occurs, or will occur*), the probability that *the third event will occur given that the first two events have occurred* (*occur, or will occur*), the probability that *the fourth event will occur given that the first three events have occurred* (*occur, or will occur*), and so forth. For three events we thus have

$$P(A \cap B \cap C) = P(A) \cdot P(B \mid A) \cdot P(C \mid A \cap B)$$

and for four events

$$P(A \cap B \cap C \cap D) = P(A) \cdot P(B \mid A) \cdot P(C \mid A \cap B) \cdot P(D \mid A \cap B \cap C)$$

where the letters can, of course, be interchanged.

To give an example, suppose we continue the illustration on page 140 and ask for the probability that if the person randomly grabs three of the mysteries, they will all be books he has not read. If we let A, B, and C denote the events that the first, second, and third books he grabs are ones he has not read, we find that $P(A) = \frac{11}{15}$, $P(B \mid A) = \frac{10}{14}$, and $P(C \mid A \cap B) = \frac{9}{13}$, and hence that the desired probability is

$$\frac{11}{15} \cdot \frac{10}{14} \cdot \frac{9}{13} = \frac{33}{91}$$

Clearly, each time he grabs a book he has not read, this reduces by 1 the number of books that are left and also the number of these that he has not yet read. A similar argument leads to the result that the probability of drawing *without replacement* four kinds in a row from an ordinary deck of 52 playing cards is

$$\frac{4}{52} \cdot \frac{3}{51} \cdot \frac{2}{50} \cdot \frac{1}{49} = \frac{1}{270{,}725}$$

To consider an example where we are not dealing with equally likely possibilities, suppose it is known from past experience that in December the probability that it will rain in a certain city on any given day is 0.60, the probability that a rainy day will be followed by a rainy day is 0.80, and the probability that a sunny day will be followed by a rainy day is 0.30. What we would like to know is the probability that *on five successive December days in that city there will be first three rainy days, then a sunny day, and then another rainy day,* assuming that the probability for rain on any given day depends only on what happened the day before. Thus, for the first day the probability is 0.60, as we do not know what happened on the day before; for the second and third days the probabilities are both 0.80, since in each case a rainy day is to follow a rainy day; for the fourth day the probability is $1-0.80=0.20$, since a sunny day is to follow a rainy day; for the fifth day the probability is 0.30, since a rainy day is to follow a sunny day; thus the answer is

$$(0.60)\cdot(0.80)\cdot(0.80)\cdot(0.20)\cdot(0.30) = 0.02304$$

In Exercise 36 on page 151 the reader will be asked to determine some other probabilities relating to this example.

EXERCISES

1. Suppose that Mr. Jones, who is not a very good card player, knows from past experience that the odds against his winning at poker are 3 to 2 while the odds against his winning at blackjack are 7 to 3. The corresponding probabilities of his winning at these games are 0.40 and 0.30, but his wife does not object to his gambling whenever they visit Las Vegas—remembering something about adding probabilities from a course she took in mathematics, she figures that the probability of his winning either at poker or at blackjack is $0.40+0.30=0.70$. Discuss the "merits" of this argument.

2. If H is the event that a job has a high starting salary and G is the event that it has a good future, state *in words* what probabilities are represented by
(a) $P(G|H)$;
(b) $P(H'|G)$;
(c) $P(H|G')$;
(d) $P(G'|H')$.

3. If I is the event that a person has a high I.Q. and W is the event that a person is socially well adjusted, express each of the following probabilities in symbolic form:

(a) the probability that a person with a high I.Q. is not well-adjusted socially;
(b) the probability that a person who is socially well adjusted does not have a high I.Q.;
(c) the probability that a person who is not well-adjusted socially also does not have a high I.Q.;
(d) the probability that a person who does not have a high I.Q. is socially well adjusted.

4. If A is the event that an astronaut is a member of the armed services, T is the event that he was once a test pilot, and W is the event that he is a well-trained scientist, express each of the following probabilities in symbolic form:
(a) the probability that an astronaut who was once a test pilot is a member of the armed services;
(b) the probability that an astronaut who is a member of the armed services is a well-trained scientist but was never a test pilot;
(c) the probability that an astronaut who is not a well-trained scientist was once a test pilot;
(d) the probability that an astronaut who is a member of the armed services but was never a test pilot is a well-trained scientist.

5. With reference to Exercise 4, express *in words* the probabilities which are represented by
(a) $P(A \mid W)$;
(b) $P(W \mid A')$;
(c) $P(A' \mid T')$;
(d) $P(A' \cap W \mid T)$;
(e) $P(A \mid W \cap T)$;
(f) $P(A \mid W \cup T)$.

6. With reference to the example on page 136, state *in words* and find the value of each of the following probabilities: (a) $P(C \mid T)$, (b) $P(T \mid C')$, and (c) $P(C' \mid T')$.

7. There are 60 qualified applicants for teaching positions in an elementary school, of which some have at least five years' teaching experience and some have not, some are married and some are single, with the exact breakdown being

	Married	*Single*
At least five years teaching experience	12	6
Less than five years teaching experience	24	18

If the order in which the applicants are interviewed by the principal is random, M is the event that the first applicant interviewed is married, and F is the event that the first applicant interviewed will have had at least five years of teaching experience, determine the following probabilities *directly* from the table:

(a) $P(M)$; (d) $P(F')$; (g) $P(F \mid M)$; (j) $P(M' \cap F')$.
(b) $P(M')$; (e) $P(M \cap F)$; (h) $P(M' \mid F')$;
(c) $P(F)$; (f) $P(M \mid F)$; (i) $P(F' \mid M')$;

8. Use the result of Exercise 7 to verify that

(a) $P(M \mid F) = \dfrac{P(M \cap F)}{P(F)}$; (c) $P(M' \mid F') = \dfrac{P(M' \cap F')}{P(F')}$;

(b) $P(F \mid M) = \dfrac{P(M \cap F)}{P(M)}$; (d) $P(F' \mid M') = \dfrac{P(M' \cap F')}{P(M')}$.

9. With reference to Exercise 7, suppose that each applicant with at least five years of teaching experience has *twice* the chance which each applicant with less than five years teaching experience has for the only opening to teach third grade. If U is the event that the third grade job will go to one of the single applicants and V is the event that it will go to an applicant with less than five years teaching experience, find each of the following probabilities:

(a) $P(U)$; (d) $P(U \cap V')$; (g) $P(U \mid V')$;
(b) $P(V)$; (e) $P(U' \cap V')$; (h) $P(V \mid U)$;
(c) $P(U \cap V)$; (f) $P(U \mid V)$; (i) $P(V' \mid U')$.

10. As a part of a promotional scheme, a department store will give a color television set to a person whose name is drawn at random from the small cards on which persons shopping at the store were asked to write their names. There is only one card per customer, and their breakdown into those who pay cash and those who charge their purchases, and those who have been shopping at the store less than a year or a year or more, is as follows:

	Pay cash	*Charge purchases*
Less than a year	216	264
A year or more	384	1056

If C stands for the event that the color television set will be won by a customer who pays cash and L stands for the event that it will be won by a customer who has been shopping at the store for less than a year, find each of the following probabilities:

(a) $P(L)$; (d) $P(L' \cap C)$; (g) $P(C \mid L)$;
(b) $P(C)$; (e) $P(L \cap C')$; (h) $P(L \mid C')$;
(c) $P(L \cap C)$; (f) $P(L \mid C)$; (i) $P(C \mid L')$.

11. With reference to Exercise 10, suppose that the department store wants to encourage new customers and fixes the drawing so that each card filled in by a customer who has been shopping at the store for less than a year has the probability $\frac{7}{4,800}$ of winning the television set, while each card filled in by a customer who has been shopping there for a year or more has the probability $\frac{1}{4,800}$. Recalculate the nine probabilities asked for in Exercise 10 under these changed conditions.

12. Referring to Exercise 8 on page 130 and its results, find
(a) the probability that a person who stops at the gas station and has his battery checked will also have his tires checked;
(b) the probability that a person who stops at the gas station and has his tires checked will also have his battery checked;
(c) the probability that a person who stops at the gas station and does not have his tires checked will have his battery checked;
(d) the probability that a person who stops at the gas station and does not have his battery checked will not have his tires checked either.

13. With reference to Exercise 10 on page 131, find
(a) the probability that a wife will vote in the local election given that her husband is going to vote;
(b) the probability that a husband will vote in the local election given that his wife is going to vote;
(c) the probability that a wife will vote in the local election given that her husband is not going to vote;
(d) the probability that a husband will not vote in the local election given that his wife is not going to vote.
(*Hint*: Draw a Venn diagram and fill in the probabilities associated with the various regions.)

14. Referring to Exercise 11 on page 131, find
(a) the probability that the secretary he hired most recently will be absent on a day when the other secretary is going to be absent;
(b) the probability that the secretary he hired most recently will be absent on a day when the other secretary is not going to be absent;
(c) the probability that the other secretary will not be absent on a day when the secretary he hired most recently is going to be absent.
(*Hint*: Draw a Venn diagram and fill in the probabilities associated with the various regions.)

15. Referring to Exercise 12 on page 131, find

(a) the probability that the teacher will get a raise given that she is going to get a promotion;
(b) the probability that the teacher will get a promotion given that she is going to get a raise;
(c) the probability that the teacher will not get a raise given that she is not going to get a promotion.

(*Hint*: Draw a Venn diagram and fill in the probabilities associated with the various regions.)

16. With reference to Exercise 7 on page 130, find
(a) $P(C \mid D)$;
(b) $P(D \mid C)$;
(c) $P(C \mid D')$;
(d) $P(D' \mid C')$.

17. In connection with the person who grabbed two paperback mysteries before getting on his plane, we showed on page 141 *by subtraction* that the probability that he will have read only one of the two mysteries is $\frac{44}{105}$. Verify this result by *adding* the probabilities of the mutually exclusive alternatives of his first grabbing the one he has read and his first grabbing the one he has not read.

18. The editor of a student newspaper wants to interview two professors chosen at random from among the eight faculty members who serve on the student publication advisory committee. If five of the faculty members are in favor of a student publication which evaluates the faculty while the others are against it, find
(a) the probability that he will get the opinions of two professors who are for the publication;
(b) the probability that he will get the opinions of two professors who are against the publication.

Also use the results of parts (a) and (b) to find
(c) the probability that he will get the opinion of one professor who is for the publication and the opinion of one professor who is against it.

[*Hint*: Use the general multiplication rule for parts (a) and (b).]

19. The probability that a bus from Phoenix to Los Angeles will leave on time is 0.80, and the probability that it will leave on time and also arrive on time is 0.72.
(a) What is the conditional probability that a bus which leaves on time will also arrive on time?
(b) If the probability that such a bus will arrive on time is 0.75, what is the conditional probability that a bus which arrives on time also left on time?

20. The probability that a student will pass a midterm examination in English is 0.60, the probability that he or she will pass the final examination in the same course is 0.72, and the probability that he or

she will pass the final examination given that he or she has passed the midterm examination is 0.90.
(a) Find the probability that a student will pass both of these examinations.
(b) If a student is known to have passed the final examination, what is the probability that he or she also passed the midterm examination?

21. Given $P(A) = 0.60$, $P(B) = 0.40$, and $P(A \cap B) = 0.24$, verify that
(a) $P(A \mid B) = P(A)$ and hence that A is independent of B;
(b) $P(B \mid A) = P(B)$ and hence that B is independent of A;
(c) $P(A \mid B') = P(A)$ and hence that A is independent of B';
(d) $P(B \mid A') = P(B)$ and hence that B is independent of A'.
(*Hint*: Draw a Venn diagram and fill in the probabilities associated with the various regions.)

22. If C and D are independent events, $P(C) = 0.60$, and $P(D) = 0.80$, find
(a) $P(C \mid D)$; (c) $P(C \cup D)$; (e) $P(C \cap D')$;
(b) $P(D \mid C)$; (d) $P(C \cap D)$; (f) $P(D' \mid C')$.

23. With reference to the two-student-interview example of Chapters 4 and 5 and the probabilities of Figure 5.2, show that *for either student* the probabilities that he will like the new record, dislike it, or will not care are, respectively, 0.50, 0.30, and 0.20. Also verify for each of the nine points of the sample space of Figure 5.2 that $P(x, y) = P(x) \cdot P(y)$, where x and y are the respective coordinates of the points.

24. With reference to Exercise 8 on page 95 and Exercise 20 on page 122, check whether events L and M are independent.

25. With reference to Exercise 11 on page 96 and Exercise 21 on page 122, check whether
(a) events Q and R are independent;
(b) events R and S are independent.

26. With reference to Exercise 14 on page 96 and Exercise 22 on page 123, check whether the event that the job is given to a married person and the event that the job is given to a native of Arizona are independent.

27. Which of the following events would you suppose are independent and which are dependent?
(a) being tired while driving and having an accident;
(b) being a teacher and having blue eyes;
(c) being born in April and having flat feet;
(d) being wealthy and collecting works of art;
(e) any two mutually exclusive events;

(f) getting 7's in two successive rolls of a pair of dice;
(g) having a flat tire and being late for work;
(h) being over twenty and smoking a pipe;
(i) being hungry and being able to afford a meal.

28. In a fifth grade class of 18 boys and 12 girls, one is chosen each week by lot to act as an assistant to the teacher. What is the probability that a girl will be chosen two weeks in a row if
(a) the same student cannot serve two weeks in a row;
(b) the restriction of part (a) is removed?

29. What is the probability of drawing two black cards in a row from an ordinary deck of 52 playing cards, of which half are black, if
(a) the drawings are without replacement;
(b) the drawings are with replacement?

30. Use the special multiplication rule for more than two independent events to find
(a) the probability of getting first two *heads* and then two *tails* in four successive flips of a balanced coin;
(b) the probability of rolling five 6's in a row with a balanced die;
(c) the probability of getting spades in the first and sixth drawing with replacement from an ordinary deck of playing cards, and some other suit in the second, third, fourth, and fifth drawings.

31. Use the special multiplication rule for more than two independent events to find
(a) the probability that a person shooting at a target will hit it three times in a row, given that the probability of his hitting the target on any one try is 0.70;
(b) the probability that four totally unrelated persons who are thirty years old will all be alive at age sixty-five, given that for any one of them the probability is 0.58;
(c) the probability that a person will answer six true-false questions correctly, if for each question the probability is 0.80 that he will know or guess the right answer.

32. What is the probability of drawing three diamonds in a row from an ordinary deck of 52 playing cards, of which 13 are diamonds, if
(a) the drawings are without replacement;
(b) the drawings are with replacement?

33. If six bullets, of which three are blanks, are randomly inserted into a gun, what is the probability that the first three bullets fired are all blanks.

34. With reference to Exercise 18 on page 148, find the probability that if the editor interviews three of the eight professors on the committee, chosen at random, all three will be in favor of the publication.

35. In a certain city, the probability of passing the test for a driver's license on the first try is 0.80, and after that the probability becomes 0.60 for each try regardless of how often a person has failed. What is the probability that a person will get his license
(a) on the third try;
(b) on the fifth try?

36. With reference to the illustration on page 144, find the probabilities that
(a) it will rain in the city on four consecutive December days;
(b) of three consecutive December days the first will be sunny and the other two will be rainy;
(c) of five consecutive December days the first will be rainy, the second sunny, the third sunny, and the fourth and fifth rainy.

37. On page 143 we indicated that it is possible for an event A to be independent of two events B and C taken individually, but not when they are taken together. Verify that this is the case in the situation pictured in Figure 6.5 by showing that $P(A \mid B) = P(A)$, $P(A \mid C) = P(A)$, but $P(A \mid B \cap C) \neq P(A)$.

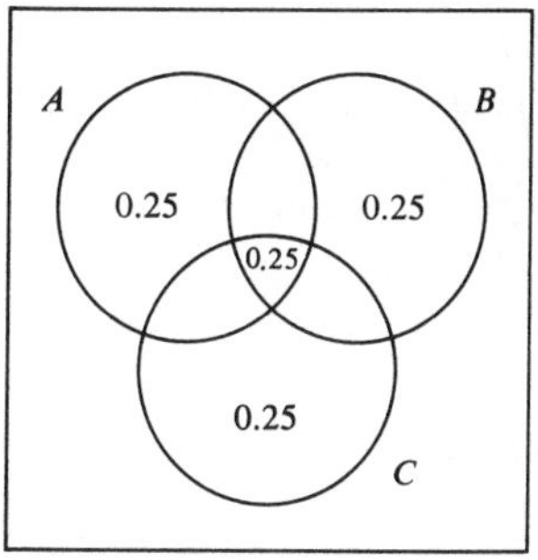

FIGURE 6.5 Venn diagram for Exercise 37.

38. It is possible for three events A, B, and C to satisfy the multiplication rule $P(A \cap B \cap C) = P(A) \cdot P(B) \cdot P(C)$, even though the events are not all *pairwise independent*. Show that this is the case in the situation pictured in Figure 6.6 by verifying that the above formula for $P(A \cap B \cap C)$ holds but that A and B are *not* independent.

39. Three events A, B, and C are independent (and this is generally used as a definition) if and only if $P(A \cap B) = P(A) \cdot P(B)$, $P(A \cap C) =$

$P(A)\cdot P(C)$, $P(B \cap C) = P(B)\cdot P(C)$, and $P(A \cap B \cap C) = P(A)\cdot P(B)\cdot P(C)$. in other words, *three events are independent when the special multiplication rule applies to any two and all three of the events.* Use this criterion to verify that the events F, G, and H of Exercise 17 on page 105 are independent when the probabilities associated with these events are as in Exercise 25 on page 123.

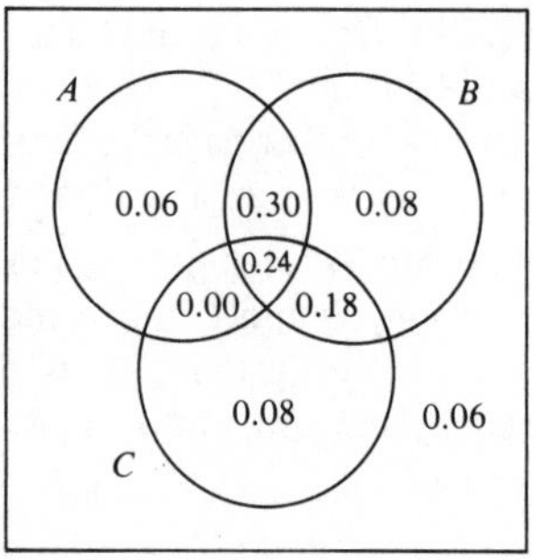

FIGURE 6.6 Venn diagram for Exercise 38.

40. Verify that the probabilities assigned to the 12 points of the sample space of Figure 6.1 are correct for a game in which we determine by lot whether to flip a balanced coin, roll a balanced die, or draw a card from a well-shuffled deck of 52 playing cards. Also verify the value given on page 134 for the probability of getting "heads, 6, or a spade."

RULE OF ELIMINATION

On page 141 we studied an example in which we were given the probability that a rare tropical disease will be diagnosed correctly and the probability that when diagnosed correctly it will be cured. The values of these probabilities were 0.25 and 0.80, respectively. Now suppose we would like to know the probability that a person having this disease will recover (or be cured, which is the same) regardless of the diagnosis. If we picture this situation as in Figure 6.7, it can be seen that the information we have is not enough—we also need the value of the probability $P(A \mid B')$ that a person having the disease will recover even though the diagnosis was not correct. Then we can calculate the probability of a patient's recovery

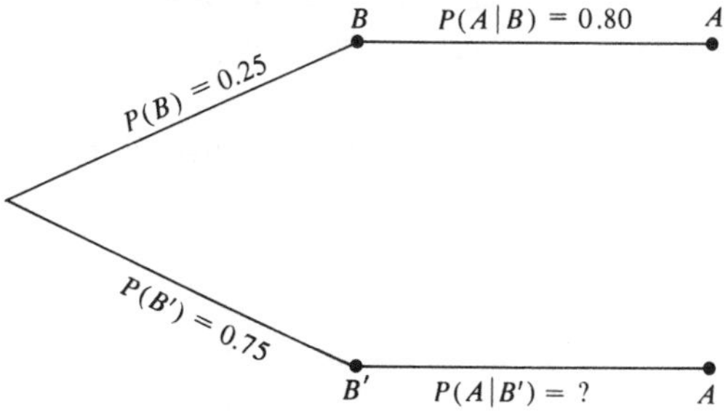

FIGURE 6.7 Tree diagram.

for each branch of the tree diagram of Figure 6.7, and *add* the results since the two possibilities (recovery via a correct diagnosis and recovery without a correct dignosis) are mutually exclusive possibilities. Thus, if the probability that a person will recover from the disease even though it was not diagnosed correctly is $P(A \mid B') = 0.12$, we find that the probabilities corresponding to the two branches of the diagram of Figure 6.7 are $P(B) \cdot P(A \mid B) = (0.25) \cdot (0.80) = 0.20$ and $P(B') \cdot P(A \mid B') = (0.75) \cdot (0.12) = 0.09$, and hence that the final answer is

$$P(A) = 0.20 + 0.09 = 0.29$$

The method we have used here is a straightforward application of the following formula which expresses the **Rule of Elimination** in its simplest form:

$$P(A) = P(B) \cdot P(A \mid B) + P(B') \cdot P(A \mid B')$$

It is used whenever some event A is "reached" via an intermediate step allowing for the two possibilities B and B'. A formal proof of the rule will be left to the reader in Exercise 11 on page 164.

To give another example where we use this rule, suppose we are concerned with the event C that the new science building of a college will be ready for use in the fall term, and the possibility E that the electricians will go on strike during the summer (before the building is completed). If the probability that the electricians will go on strike is 0.60, and the probability that the building will be completed in time is 0.15 or 0.75, respectively, depending on whether or not the electricians will go on strike, we find that the probability that the electricians will not go on strike is $1 - 0.60 = 0.40$ and, hence, that

$$\begin{aligned} P(C) &= P(E)\cdot P(C \mid E) + P(E')\cdot P(C \mid E') \\ &= (0.60)\cdot(0.15) + (0.40)\cdot(0.75) \\ &= 0.39 \end{aligned}$$

This is the desired probability that the building will be completed on time.

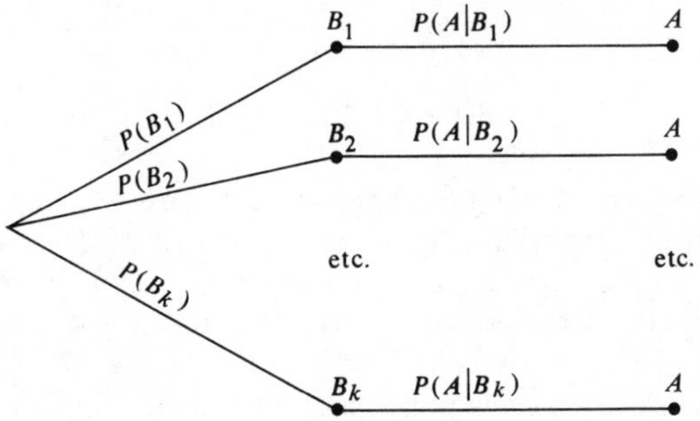

FIGURE 6.8 Tree diagram for Rule of Elimination.

The rule of elimination *in general* pertains to situations like the one pictured in Figure 6.8, where the final outcome A is "reached" via k different possibilities $B_1, B_2, \ldots,$ and B_k. As before, the probability that A is "reached" via B_1 is $P(B_1)\cdot P(A \mid B_1)$, the probability that it is "reached" via B_2 is $P(B_2)\cdot P(A \mid B_2)$, ... and the probability that it is "reached" via B_k is $P(B_k)\cdot P(A \mid B_k)$, and we get the probability of A by adding all these probabilities *provided the B's are mutually exclusive and account for all the possibilities.* Thus, we arrive at the following result which expresses the **Rule of Elimination** in its general form:

If $B_1, B_2, \ldots,$ and B_k constitute a set of mutually exclusive events of which one must occur, then for any event A

$$\begin{aligned} P(A) = P(B_1)\cdot P(A \mid B_1) &+ P(B_2)\cdot P(A \mid B_2) \\ &+ \ldots + P(B_k)\cdot P(A \mid B_k) \end{aligned}$$

Note that for $k = 2$ the two possibilities B_1 and B_2 can be written as B and B', and the formula reduces to the one on page 153.

In the $\sum$ notation, with which the reader may be familiar from previous studies in mathematics and which is simply a form of *mathematical shorthand*, the above formula can be written as

$$P(A) = \sum_{i=1}^{k} P(B_i)\cdot P(A \mid B_i)$$

Here $\sum$ (capital *sigma,* the Greek capital S) stands for "sum, and $\sum_{i=1}^{k} P(B_i)\cdot P(A \mid B_i)$ tells us that we must *add* the quantities $P(B_i)\cdot P(A \mid B_i)$, with i first taking on the value 1, then the value 2, then 3,..., and finally the value k. Exercises 23 and 24 on page 166 are designed to give the reader some experience with this notation.

To illustrate the general form of the rule of elimination, suppose that three candidates, Mr. Carter, Mr. Dennis, and Mr. Evans, are running for the office of governor of a certain state. The biggest campaign issue is that of lowering property taxes, and it can be said that the probabilities for lower property taxes are 0.80, 0.10, and 0.55, depending on whether Mr. Carter, Mr. Dennis, or Mr. Evans is elected. What we would like to know is the probability that property taxes will be lowered *regardless of who is elected,* and to this end we will have to know something about the three candidates' chances of winning the election. Thus, let us suppose that the odds favor Mr. Carter by 3 to 2, but are 4 to 1 against Mr. Dennis as well as Mr. Evans. (These odds are consistent since the corresponding probabilities are 0.60, 0.20, and 0.20, which add up to 1.) Now, if we let A denote the event that property taxes will be lowered, and B_1, B_2, and B_3, respectively, denote the events that Mr. Carter, Mr. Dennis, or Mr. Evans is elected, the information we have can be written as $P(B_1) = 0.60$, $P(B_2) = 0.20$, $P(B_3) = 0.20$, $P(A \mid B_1) = 0.80$, $P(A \mid B_2) = 0.10$, and $P(A \mid B_3) = 0.55$, and substitution into the formula yields

$$\begin{aligned} P(A) &= (0.60)\cdot(0.80) + (0.20)\cdot(0.10) + (0.20)\cdot(0.55) \\ &= 0.61 \end{aligned}$$

To really *understand* what is going on, it is best to picture the whole situation by means of a tree diagram like that of Figure 6.8. Here the probabilities associated with the three branches are $(0.60)(0.80) = 0.48$, $(0.20)(0.10) = 0.02$, and $(0.20)(0.55) = 0.11$, and the answer is given by the sum of these three probabilities, namely $0.48 + 0.02 + 0.11 = 0.61$. The formal proof of the rule of elimination is very similar to the one which the reader will be asked to give in Exercise 11 on page 164 for the case where $k = 2$; it is not difficult, but we shall not go into it in this book.

The method we have presented here for *one* intermediate step can easily be generalized so that it applies to situations in which there are *several* intermediate steps. Here, too, it is possible to write down general formulas, but it is easier (and, in fact, *more instructive*) to work with tree diagrams. Suppose, for instance, that we want to determine the probability that a home airconditioning system will function efficiently on the basis of the following information: The probabilities that the system, itself, is in excellent condition, good condition, or poor condition are, respectively, 0.60, 0.30, and 0.10; the probability that the system will be installed properly is 0.90, and the probabilities that the system will function efficiently under the various conditions are as indicated in the third step of the tree diagram of Figure 6.9. For instance, if the system is in excellent condition. and properly installed, the probability that it will work efficiently is 0.90; if the system is in excellent condition but not properly installed, the probability that it will work efficiently is 0.75; if the system is in good condition and properly installed, the probability that it will work efficiently is 0.60; and if the system is in poor condition and improperly installed,

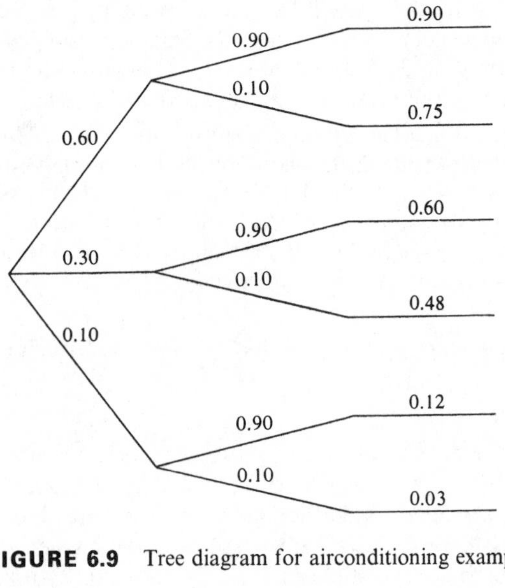

FIGURE 6.9 Tree diagram for airconditioning example.

the probability that it will function efficiently is 0.03. Thus, the probability associated with the first path along the branches of the tree diagram is $(0.60)(0.90)(0.90) = 0.486$, the one associated with the second path is $(0.60)(0.10)(0.75) = 0.045$, and those associated with the third, fourth, fifth, and sixth paths are, respectively, $(0.30)(0.90)(0.60) = 0.162$, $(0.30)(0.10)(0.48)$

$= 0.0144$, $(0.10)(0.90)(0.12) = 0.0108$, and $(0.10)(0.10)(0.03) = 0.0003$. Since these six possibilities are mutually exclusive, we can again add the probabilities associated with the various paths and, thus, arrive at the result that the probability that the airconditioning system will work efficiently is

$$0.486 + 0.045 + 0.162 + 0.0144 + 0.0108 + 0.0003 = 0.7185$$

or approximately 0.72.

BAYES' RULE

Although the symbols $P(A \mid B)$ and $P(B \mid A)$ may look alike, there is a great difference between the corresponding probabilities. As we saw on page 136, $P(T \mid C)$ was the probability that the editor will select a letter which favors the teacher *given that it was written by a student,* and the correct answer to part (a) of Exercise 6 on page 145 is that $P(C \mid T)$ is the probability that the editor will select a letter which was written by a student *given that it favors the teacher.* The important distinction between these two probabilities is that they pertain to *different sample spaces*—for $P(T \mid C)$ the selection is limited to letters written by students, and for $P(C \mid T)$ the selection is limited to letters favoring the teacher. To consider another example, suppose that R is the event that someone has committed a robbery and G is the event that he will be judged guilty. Then, $P(G \mid R)$ is the probability that *the person who committed the robbery will be found guilty,* and $P(R \mid G)$ is the probability that *the person who is found guilty actually did commit the robbery.*

Since there are many problems which involve such *matched pairs* of conditional probabilities, let us try to find a formula which expresses the relationship between $P(A \mid B)$ and $P(B \mid A)$ for any two events A and B. Surprisingly, this is easily done—all we have to do is equate the expressions given for $P(A \cap B)$ in the two forms of the general multiplication rule on pages 139 and 140, and we get

$$P(A) \cdot P(B \mid A) = P(B) \cdot P(A \mid B)$$

Then, if we divide the expressions on both sides of this equation by $P(A)$, we get

$$P(B \mid A) = \frac{P(B) \cdot P(A \mid B)}{P(A)}$$

and we now have a formula which expresses one of these conditional probabilities in terms of the other. To illustrate its use, suppose that in the situation described on page 153 it turns out that the new science building is completed on time, and that a professor (who discovers this upon returning from a summer job at another college) wonders whether the electricians had gone out on strike. Of course, he can ask, but before he finds out for sure he can use the figures on page 154, namely, $P(E) = 0.60$, $P(C \mid E) = 0.15$, and $P(C) = 0.39$, and conclude that

$$P(E \mid C) = \frac{P(E) \cdot P(C \mid E)}{P(C)} = \frac{(0.60)(0.15)}{0.39} = \frac{0.09}{0.39} = \frac{3}{13}$$

or approximately 0.23. This is the probability that the electricians had their strike *given that the building is known to have been completed on time.*

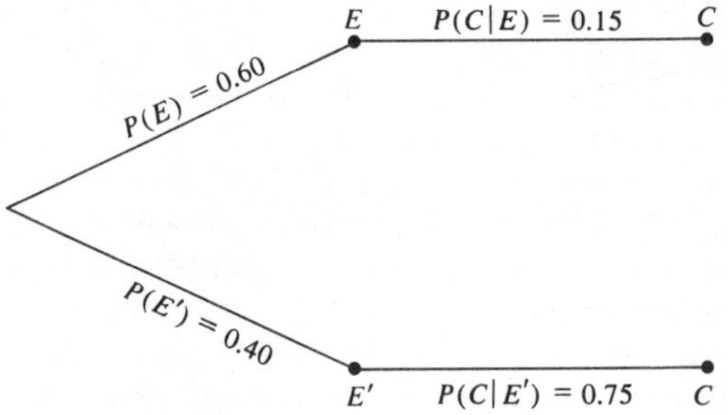

FIGURE 6.10 Tree diagram.

Since formulas are easy to forget, it is helpful in problems like this to analyze the situation by means of a tree diagram like that of Figure 6.6. With reference to Figure 6.10 we can thus say that in the *numerator* of the formula for $P(E \mid C)$ we put the probability associated with the first branch of the tree diagram, namely, the one which represents completion of the building via event E, and in the *denominator* we put the *sum* of the probabilities associated with both branches of the diagram. [As we saw in the preceding section, this sum gives the probability $P(C)$ of completing the building *with or without the strike.*] Once we have found $P(E \mid C)$, the value of $P(E' \mid C)$ can be obtained by subtraction from 1, and it is $1 - \frac{3}{13} = \frac{10}{13}$, but with reference to Figure 6.10 we can argue instead that $P(E' \mid C)$ is given by the *quotient* of the probability associated with the second branch of the tree diagram, the one leading to the completion of the building via event E', divided by the sum of the probabilities associated with both branches, namely, by

$$P(E' \mid C) = \frac{P(E')\cdot P(C \mid E')}{P(E)\cdot P(C \mid E) + P(E')\cdot P(C \mid E')}$$

$$= \frac{(0.40)(0.75)}{(0.60)(0.15) + (0.40)(0.75)}$$

$$= \frac{0.30}{0.39} = \frac{10}{13}$$

The formula which we used in the last example is **Bayes' rule** in its simplest form. This rule, which we shall study in the next few pages in some detail, is named after the English clergyman Thomas Bayes, and it dates back to the second half of the eighteenth century. Although there is no question about the *validity* of this rule, questions have been raised about its *applicability*, and this is due to the fact that it involves a "backward" or "inverse" sort of reasoning, namely, reasoning *from effect to cause*. For instance, the high value which we obtained for $P(E' \mid C)$ in the preceding paragraph reflects the idea that the completion of the building was "caused," or brought about, by there not having been a strike. Similarly, in the example on page 153, a high value of $P(B \mid A)$ would reflect the idea that a patient's recovery from the disease was "caused" by a correct diagnosis, and a high value of $P(B' \mid A')$ would reflect the idea that a patient's failure to recover was "caused" by the absence of a correct diagnosis; as the reader will be asked to verify in Exercise 17 on page 165, the values of these two probabilities are, respectively, $\frac{20}{29}$ and $\frac{66}{71}$.

In the preceding section we generalized the rule of elimination by considering intermediate steps consisting of the k alternatives $B_1, B_2, \ldots,$ and B_k instead of the two possibilities B and B'. If we do the same with regard to Bayes' rule and refer to the tree diagram of Figure 6.8, it suggests itself that the probability $P(B_1 \mid A)$ may well be given by the *quotient* of the probability associated with the first branch, the one which leads to A via B_1, divided by the *sum* of the probabilities associated with all the branches; that the probability $P(B_2 \mid A)$ may well be given by the *quotient* of the probability associated with the second branch, the one which leads to A via B_2, divided by the *sum* of the probabilities associated with all the branches; and so on. This is, indeed, the case, and when the B's are mutually exclusive events of which one must occur, **Bayes' rule in its general form** can be written as*

* Note that we are using the subscript r in $P(B_r \mid A)$ since the subscript i is already used in the sum in the denominator, and it would be very confusing to use the same letter in two different connotations.

$$P(B_r | A) = \frac{P(B_r) \cdot P(A | B_r)}{\sum_{i=1}^{k} P(B_i) \cdot P(A | B_i)} \quad \textbf{for } r = 1, 2, \ldots, \textbf{ or } k$$

Note that the *numerator* of this formula for $P(B_r | A)$ is the probability associated with the rth branch of the tree diagram of Figure 6.8, the one leading to A via B_r, and the *denominator* consists of the *sum* of the probabilities associated with all the branches (given in the $\sum$ notation explained on page 155). This result is quite easy to prove. Substituting B_r for B into the formula on page 157 (which, itself, was an immediate consequence of the general multiplication rules on pages 139 and 140), we get

$$P(B_r | A) = \frac{P(B_r) \cdot P(A | B_r)}{P(A)}$$

and the only thing that remains to be done is to substitute for $P(A)$ the sum $\sum_{i=1}^{k} P(B_i) \cdot P(A | B_i)$ in accordance with the rule of elimination. When it comes to practical applications, the reader should find it easiest to think of Bayes' rule as it was introduced first, namely, in terms of the probabilities associated with the various branches of a tree diagram like that of Figure 6.8.

To illustrate Bayes' rule in its general form, let us return to the example on page 155, and let us suppose that a person who has moved away from the state before the gubernatorial election discovers that taxes on property he owns in that state have been lowered. What he would like to know is the probability that this pleasant development was brought about by the election of Mr. Carter. Using the notation introduced on page 155 and the values assigned there to the various probabilities, we find that the answer to his question is given by

$$\begin{aligned} P(B_1 | A) &= \frac{P(B_1) \cdot P(A | B_1)}{P(B_1) \cdot P(A | B_1) + P(B_2) \cdot P(A | B_2) + P(B_3) \cdot P(A | B_3)} \\ &= \frac{(0.60)(0.80)}{(0.60)(0.80) + (0.20)(0.10) + (0.20)(0.55)} \\ &= \frac{0.48}{0.61} \end{aligned}$$

which is approximately 0.79. Actually, we could have saved ourselves a good deal of work since we already showed on page 155 that the sum of the probabilities in the denominator equals $P(A) = 0.61$. Furthermore, we showed that the probabilities associated with the three branches of a tree diagram like that of Figure 6.8 are, respectively, 0.48, 0.02, and 0.11, so that we could have argued *directly* that the probability $P(B_1 | A)$ is

$$\frac{0.48}{0.48+0.02+0.11} = \frac{0.48}{0.61}$$

The same method leads to probabilities of $\frac{0.02}{0.61}$ and $\frac{0.11}{0.61}$, or approximately 0.03 and 0.18, that the reduction of property taxes was brought about by the election of Mr. Dennis or the election of Mr. Evans.

To give another example, suppose that a department store employs four gift wrappers at Christmas time. Betty, who works long hours, wraps 36 percent of all packages and fails to remove the price tag 2 percent of the time, and Susan who also works long hours, wraps 40 percent of all packages and fails to remove the price tag 3 percent of the time. The other two, Carol and Jean, work relatively shorter hours; Carol wraps 10 percent of all packages and fails to remove the price tag 6 percent of the time, and Jean wraps 14 percent of all packages and fails to remove the price tag 8 percent of the time. What we would like to know is *the probability that a gift bought at this store which did not have its price tag removed was wrapped by Betty.* Picturing this situation by means of the tree diagram of Figure 6.11, we find that the probabilities associated with

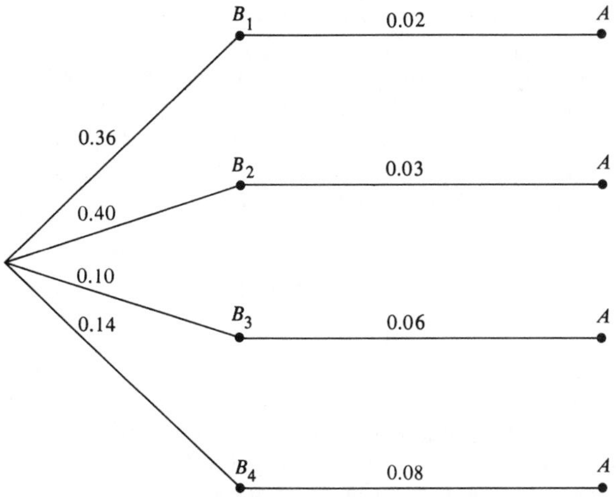

FIGURE 6.11 Tree diagram for gift = wrapping example.

the four branches are, respectively, $(0.36)(0.02) = 0.0072$, $(0.40)(0.03) = 0.0120$, $(0.10)(0.06) = 0.0060$, and $(0.14)(0.08) = 0.0112$, and, hence, that the desired probability is

$$P(B_1 \mid A) = \frac{0.0072}{0.0072+0.0120+0.0060+0.0112} = \frac{0.0072}{0.0364}$$

which is approximately 0.20. Thus, the odds are just about 4 to 1 that Betty did *not* wrap the gift with the price tag left on even though she is known to wrap 36 per cent of all gifts. For Jean the situation is reversed; the probability that she is responsible for having wrapped a gift is *increased* from 0.14 to

$$P(B_4 \mid A) = \frac{0.0112}{0.0364}$$

or approximately 0.31, once it is known that the price tag has been left on. Of course, the reason for this is her very high percentage of mistakes.

In connection with problems like these, we often refer to the probabilities $P(B_1)$, $P(B_2)$, ..., and $P(B_k)$ as the **prior probabilities** of events B_1, B_2, ..., and B_k, which may be looked upon as possible "causes" of event A. Correspondingly, we refer to the probabilities $P(B_1 \mid A)$, $P(B_2 \mid A)$, ..., and $P(B_k \mid A)$ as the **posterior probabilities** of these "causes," given that event A has occurred (occurs, or will occur). Note that we have used the word "cause" in quotes as we are trying to imply that the B's lead to A *without there having to be a strict cause-and-effect relationship.* Thus, in the last example, the prior probability that a gift was wrapped by Betty is 0.36, but once we know that the price tag has been left on, the posterior probability that the gift was wrapped by her is only 0.20. This illustrates how the probability of an event can be affected by new information, and this is very important in statistics, where we add to our information by gathering data. Further examples dealing with this problem will be given in Chapter 7.

EXERCISES

1. Ray has enrolled as a freshman at an Eastern university and the probability that he will get a scholarship is 0.35. If he gets a scholarship the probability that he will graduate is 0.82, and if he does not get a scholarship the probability that he will graduate is only 0.44. What is the probability that he will graduate?

2. In a T-maze, a rat is given food if it turns left and an electric shock if it turns right. On the first trial there is a fifty-fifty chance that a rat will turn either way; then, it if receives food on a given trial the probability that it will turn left on the next trial is 0.72, and if it receives a shock on a given trial the probability that it will turn left on the next trial is 0.84.

(a) What is the probability that a rat will turn left on the second trial?

(b) Assuming that the probability of a rat turning left on any given trial depends only on what it did in the preceding trial, find the probability that a rat will turn left on the third trial. (*Hint*: draw a tree diagram and calculate, and then add, the probabilities associated with the four paths which lead to the rat turning left on the third trial.)

3. With reference to Exercise 19 on page 148, suppose that the probability that a bus which leaves Phoenix late will arrive on time in Los Angeles is 0.78. What is the probability that any one of these buses from Phoenix to Los Angeles will arrive on time?

4. The odds are 3 to 2 that a famous European distance runner will enter the Boston marathon. If he does not enter, the probability that the United States champion will win is 0.66, but if he enters, the probability that the United States champion will win is only 0.18. What is the probability that the United States champion will win the race?

5. With reference to the illustration on page 144, show that the probability is 0.60 that it will rain on the *second* day of a December vacation which a person is spending in the given city. *Should we be surprised that this value equals the probability that it will rain there on any day in December?* Verify that the probability is also 0.60 for the *third* day of a vacation which a person is spending in the given city some time in December.

6. In an automobile assembly plant, assembly lines A, B, and C account respectively, for 45 percent, 30 percent, and 25 percent of the plant's total output. If the probability is 0.004 that a car coming off assembly line A has a defect due to poor workmanship, while the corresponding probabilities for assembly lines B and C are, respectively, 0.006 and 0.010, what is the probability that a car assembled at this plant will have a defect due to poor workmanship?

7. The probabilities that a cosmetics manufacturer's vice president in charge of advertising will decide to promote a new shaving lotion on television during football games, a panel show, a soap opera, or the evening news, are, respectively, 0.45, 0.30, 0.10, and 0.15. If he decides on the football games, he figures, the probability that they will get a high rating is 0.62; on the other hand, if he decides on a panel show, soap opera, or the evening news, the corresponding probabilities are, respectively, 0.35, 0.26, and 0.12. According to these figures, what is the probability that the show on which the new shaving lotion will be promoted will get a high rating?

8. In Exercise 4 on page 11 we considered the situation where the National League team was leading the American League team 2 games to 1 in a (best of seven) World series. Assuming that the two teams are evenly matched, what is the probability that
(a) there will be a seventh game;
(b) the American League team will win the series?
(*Hint*: Refer to the tree diagram constructed in that exercise and add the probabilities associated with the appropriate paths.)

9. Referring to Exercise 7 on page 12, suppose that the probabilities are, respectively, 0.80, 0.15, and 0.05 that the golf pro can sell zero, one, or two sets of the clubs on any given day, and that the demand will not carry over from one day to the next if he does not have enough sets in stock. Use the tree diagram constructed in that exercise to find the probabilities that on the first two days of that week the pro will sell
(a) altogether one set of these clubs;
(b) at least three sets of these clubs.

10. With reference to Exercise 8 on page 12, suppose that the probabilities are, respectively, 0.50, 0.40, and 0.10 that the student will study zero, one, or two hours on any given night, and that the number of hours he studies on different nights are independent. What is the probability that he will study altogether six hours for the examination on four consecutive nights?

11. Prove the special form of the rule of elimination given on page 153 by making use of part (c) of Exercise 14 on page 104 and the fact that the events $A \cap B$ and $A \cap B'$ are mutually exclusive.

12. With reference to Exercise 1, suppose that years later we hear that Ray graduated from the given university. What is the probability that he did get the scholarship?

13. With reference to part (a) of Exercise 2, suppose that a rat turns left on the second trial. What is the probability that it had turned left on the first trial?

14. With reference to Exercise 3 above and Exercise 19 on page 148, what is the probability that a bus from Phoenix to Los Angeles which arrives on time, had left on time?

15. With reference to Exercise 4, suppose we hear that the U.S. champion won the race. Not having any details about what happened, what odds should we give that the famous European distance runner did not enter?

16. (From Hans Reichenbach's *The Theory of Probability*, University of California Press, 1949.) Mr. Smith's gardener is not dependable; the probability that he will forget to water the rosebush during Smith's absence is $\frac{2}{3}$. The rosebush is in questionable condition anyhow; if watered, the probability for its withering is $\frac{1}{2}$; if it is not watered, the probability for its withering is $\frac{3}{4}$. Upon returning Smith finds that the rosebush has withered. What is the probability that the gardener did not water the rosebush?

17. Verify the values given on page 159 for the probabilities $P(B|A)$ and $P(B'|A')$ that a patient's cure from the rare disease was brought about by a correct diagnosis and that a patient's failure to recover was due to the lack of a correct diagnosis.

18. A history teacher tells his class that the odds are 4 to 1 that a student who regularly does his homework will get a passing grade, but 3 to 1 *against* a student who does not regularly do his homework. Also, he figures that 60 percent of his students regularly do their homework. What is the probability that a student who gets a passing grade regularly did his homework?

19. With reference to Exercise 7, suppose we find out later that the television program in which the new shaving lotion was advertised received a high rating. What is the probability that this success was brought about (but not necessarily "caused by") by the vice president's decision to advertise this new product on telecasts of football games?

20. With reference to Exercise 6, what is the probability that a car assembled at this plant which has a defect due to poor workmanship came off assembly line C? What are the corresponding probabilities for assembly lines A and B?

21. With reference to Exercise 8, suppose we hear that the American League team won the series. What is the probability that it took seven games?

22. Arriving at a New Year's Eve party, Mr. Jones and his wife are offered to help themselves from a plate on which there are six donuts. Mr. Jones knows that his host is a practical joker and suspects that some, or all, of the donuts may be filled with mustard instead of jelly. Not having any direct information, he assigns a prior probability of $\frac{1}{7}$ to each of the possibilities that 0, 1, 2, ..., or all six of the donuts are filled with mustard. Letting his wife go first and randomly pick one of the donuts, he finds that it is perfectly alright—no mustard. With this information, would 5 to 2 be fair odds that at least one of the donuts is filled with mustard?

23. Write each of the following in full, that is, without a summation sign:

(a) $\sum_{i=1}^{6} x_i$;

(b) $\sum_{i=1}^{4} P(A_i)$;

(c) $\sum_{i=1}^{n} a_i \cdot p_i$;

(d) $\sum_{i=1}^{5} P(B_i|A)$;

(e) $\sum_{j=1}^{3} P(A \cap B_j)$;

(f) $\sum_{j=2}^{4} (1-p_j)$.

24. Write each of the following sums in the $\sum$ notation:

(a) $y_1+y_2+y_3+y_4+y_5+y_6+y_7+y_8$;

(b) $(x_1-2)+(x_2-2)+(x_3-2)+(x_4-2)$;

(c) $P(A_1 \cup B_1)+P(A_2 \cup B_2)+P(A_3 \cup B_3)$;

(d) $P(C'_1)+P(C'_2)+ \dots +P(C'_k)$;

(e) $P(A_1 \cap B_1)+P(A_1 \cap B_2)+P(A_1 \cap B_3)+P(A_1 \cap B_4)$;

(f) $P(D|C_3)+P(D|C_4)+P(D|C_5)+P(D|C_6)$.

7

PROBABILITY FUNCTIONS

INTRODUCTION

In most probability problems we are interested only in one or in a few aspects of the possible outcomes. In the game of craps, for example, we are interested only in the total which a player rolls with a pair of dice, and a player who rolls a 1 and a 6 is just as happy (or unhappy, as the case may be) as a player who rolls a 2 and a 5 or a 3 and a 4. Similarly, if a student takes a true-false test consisting, say, of 100 questions, his grade will depend only on the number of questions he misses, and it does not matter whether he misses the fourteenth, seventeenth, fifty-fifth, and eighty-first questions, or the twenty-sixth, forty-third, seventy-ninth, and ninety-fifth. To give an example where *two or more* aspects of the outcomes are of interest, consider an inspector examining the quality of television tubes—he may be interested in their durability and brightness (and, perhaps, also in their price and size). Also, a market research worker may ask housewives about the size of their family, their husband's income, and how many times they have tried a certain product (and, perhaps, also about their age, their dress size, and the number of years they have gone to school).

In each of these examples we are interested in a number, or numbers, that are associated wth the outcomes of situations involving an element of chance, namely, in the values taken on by so-called **random variables**. To be more explicit, let us consider Figure 7.1, which (like Figure 5.2 on page 118) pictures the sample space for the two-student-interview example, with a probability attached to each possible outcome. Note, however, that in contrast to Figure 5.2, we have attached another number to each point: The number 2 to the point (1, 1), the number 1 to the points (1, 2), (1, 3), (2, 1), and (3, 1), and the number 0 to the other four points. As should be apparent, we have thus associated with each point of the sample space the corresponding number of students who like the new record, namely, the value of the random variable "the number of students who like the new record." Since "associating a number with each point of a sample space" is just another way of saying that we are defining a *function* over the points of the sample space, random variables are, strictly speaking, functions. Since this terminology is confusing, however, it will be convenient to think of random variables rather as quantities which can take on different values depending on chance. Thus, the price of a share of IBM stock is a random variable, and so is the crowd at a football game, the annual

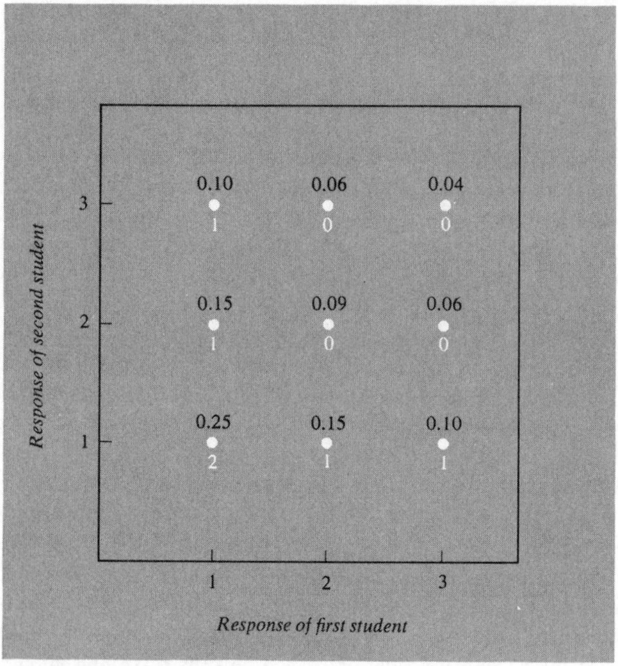

FIGURE 7.1 Sample space showing values of random variable.

production of wheat in the United States, the number of speeding tickets issued each day on the Pasadena freeway, the wind velocity at Kennedy airport, the number of mistakes a student makes in a quiz, and so forth.

PROBABILITY FUNCTIONS

In the study of random variables, we are usually interested mainly in the probabilities with which they take on the various values within their range. Thus, in connection with Figure 7.1 we may be interested mainly in the values shown in the following table, which were obtained by adding the probabilities associated with the respective points:

Number of students who like the new record	*Probability*
0	$0.06+0.04+0.09+0.06 = 0.25$
1	$0.15+0.10+0.15+0.10 = 0.50$
2	0.25

It is important to note that, depending on one's interest, any number of random variables can be defined over the same sample space. Had we been interested in "the number of students who dislike the new record" or "the number of students who like the new record *minus* the number of students who dislike it," we would have obtained the following two tables, which the reader will be asked to check in Exercise 1 on page 172 (see also the hint given in that exercise):

Number of students who dislike the new record	*Probability*
0	0.49
1	0.42
2	0.09

Number of students who like the new record minus number of students who dislike it	*Probability*
−2	0.09
−1	0.12
0	0.34
1	0.20
2	0.25

All these tables serve to illustrate what we mean by a **probability function**, namely, *a correspondence which assigns probabilities to the values of a random variable.* To give another example, let us find the probability function for the number of heads in three flips of a balanced coin. Letting H stand for *heads* and T for *tails,* we can list the eight possibilities as

HHH HHT HTH THH HTT THT TTH TTT

and since each has the probability $\frac{1}{2}\cdot\frac{1}{2}\cdot\frac{1}{2} = \frac{1}{8}$, we arrive at the probabilities shown in the following table:

Number of heads	*Probability*
0	$\frac{1}{8}$
1	$\frac{3}{8}$
2	$\frac{3}{8}$
3	$\frac{1}{8}$

Although this last example may seem trivial and of interest mainly to someone addicted to flipping coins, it can provide answers to questions relating to entirely different matters. Suppose, for example, that a home builder wants to develop a tract of land by building 400 houses meant essentially for low-income families with three small children. What he wants to know is *how many of these houses should have two bedrooms* (*one for the children*) *and how many should have three bedrooms* (*two for the children*). He assumes that all children of the same sex will share a bedroom, and that there is a fifty-fifty chance for each child to be a boy or a girl (which is not quite correct, but close). Referring to the above table of the probability function for the number of heads in three flips of a balanced coin (which is now the number of boys in a family with three children), we find that there are either three boys or three girls with the probability $\frac{1}{8}+\frac{1}{8}=\frac{1}{4}$, and at least one boy and one girl with the probability $\frac{3}{8}+\frac{3}{8}=\frac{3}{4}$. Translating these results into the corresponding need for two- and three-bedroom homes, we find that

Number of bedrooms needed	*Probability*
2	0.25
3	0.75

and it follows that the builder should construct 400(0.25) = 100 two-bedroom homes and 400(0.75) = 300 three-bedroom homes.

Whenever possible, we try to express probability functions by means of formulas (equations) which enable us to calculate the probabilities associated with the various values of a random variable. For instance, using the functional notation, we can write*

$$f(x) = \frac{\binom{2}{x}}{4} \quad \text{for } x = 0, 1, \text{ or } 2$$

for the number of students who like the new record in the two-student-interview example. Here $f(0)$, $f(1)$, and $f(2)$ denote the probabilities that the random variable (the number of students who like the new record) will take on the values 0, 1, or 2, and since $\binom{2}{0} = 1$, $\binom{2}{1} = 2$, and $\binom{2}{2} = 1$ in

* Most of the time we shall write the probability that a random variable takes on the value x as $f(x)$, but we could just as well write it as $g(x)$, $h(x)$, $b(x)$, and so on.

accordance with the definition of these combinatorial symbols in Chapter 1, we find that

$$f(0) = \frac{\binom{2}{0}}{4} = \frac{1}{4}, \quad f(1) = \frac{\binom{2}{1}}{4} = \frac{2}{4}, \quad \text{and} \quad f(2) = \frac{\binom{2}{2}}{4} = \frac{1}{4}$$

This agrees with the values given in the table on page 169.

In connection with probability functions it should be noted that (1) *their values, being probabilities, must be real numbers on the interval from 0 to 1,* and (2) *the sum of all the values of a probability function must always equal 1.* This second property follows from the fact that "a random variable taking on different values" are mutually exclusive events of which one must occur. Although a rather fine distinction is made in some books between probability functions and **probability distributions** (or simply **distributions**), we shall not do so here. Indeed, a probability function tells us how the total probability of 1 for the entire sample space is *distributed* among the possible values which a random variable can take on.

THE MEAN

Since the table on page 169 shows that the probabilities are 0.25, 0.50, and 0.25 that zero, one, or two of the students will like the new record, we can say that when two students are thus interviewed, we can *expect*

$$0(0.25)+1(0.50)+2(0.25) = 1$$

of the two to like the new record. Of course, this is a *mathematical expectation* which was obtained in accordance with the formula on page 62—it is the sum of the products obtained by multiplying each value of the random variable by the corresponding probability. In fact, we refer to the result as the **expected value of the random** variable; in general, if a random variable takes on the values x_1, x_2, ..., and x_k with the probabilities $f(x_1), f(x_2), \ldots,$ and $f(x_k)$, its expected value is given by

$$x_1 \cdot f(x_1)+x_2 \cdot f(x_2)+ \ldots +x_k \cdot f(x_k)$$

This sum is also referred to as the **mean** of the distribution of the random variable and it is usually denoted with the Greek letter μ (*mu*). Using the $\sum$ notation which we explained on page 155, we can thus write

$$\mu = \sum_{i=1}^{k} x_i \cdot f(x_i)$$

As it is used here, the term "mean" comes from statistics, where it designates what is colloquially referred to as an "average."

To give a few examples, let us find the mean of the distribution of the number of heads in three flips of a balanced coin, and the mean of the distribution of the number of mysteries the person will have read in the example on page 140. Referring to the table on page 169, we find that

$$\mu = 0(\tfrac{1}{8}) + 1(\tfrac{3}{8}) + 2(\tfrac{3}{8}) + 3(\tfrac{1}{8}) = 1\tfrac{1}{2}$$

for the number of heads in three flips of a balanced coin. Of course, we cannot actually get $1\frac{1}{2}$ heads in three flips of a coin, and it should be understood that a mean is always an "average" in the sense of a mathematical expectation. So far as the other example is concerned, we know from page 141 that the probabilities are $\frac{11}{21}$, $\frac{44}{105}$, and $\frac{2}{35}$ that the person will have read zero, one, or two of the mysteries, so that we can expect that he will have read

$$\mu = 0 \cdot \tfrac{11}{21} + 1 \cdot \tfrac{44}{105} + 2 \cdot \tfrac{2}{35} = \tfrac{8}{15}$$

of the mysteries.

EXERCISES

1. Verify the values of the probability functions given on page 169 for
(a) the number of students who dislike the new record;
(b) The number of students who like the new record minus the number of students who dislike it. [*Hint*: Refer to Figure 7.1, find the value of the random variable for each of the nine points, and then add the respective probabilities. For instance, the value of the random variable is -2 only for the point $(2,2)$ so that the corresponding probability is 0.09, and the value of the random variable is -1 for the points $(2,3)$ and $(3,2)$ so that the corresponding probability is $0.06+0.06 = 0.12$.]

2. Figure 4.6 pictures the sample space for the example where a baker has four chocolate cream pies which have to be sold either on Friday, the day they are baked, or on Saturday. Assuming that each point of the sample space has the probability $\frac{1}{15}$, construct tables showing the values of the probability functions of the following random variables:

(a) the number of chocolate cream pies which he sells on Friday;
(b) the total number of chocolate cream pies which he sells either on Friday or Saturday;
(c) the number of chocolate cream pies that remain unsold.
(*Hint*: In each case, attach the value of the random variable and the probability to each point of the sample space, and then add the probabilities of the points which correspond to the respective values of the random variable.)

3. Referring to the restaurant of Exercise 11 on page 96 and assuming that each of the 20 points of the sample space has the probability $\frac{1}{20}$, construct tables showing the values of the probability functions of the following random variables:
(a) the number of tables in the small room that are being used;
(b) the number of tables in the large room that are being used;
(c) the total number of tables that are being used.
(*Hint*: In each case, attach the value of the random variable and the probability to each point of the sample space, and then add the probabilities of the points which correspond to the respective values of the random variable.)

4. Referring to the taxicab example of Exercise 8 on page 95 and the probabilities assigned to the 15 points of the sample space in Exercise 20 on page 122, construct a table showing the values of the probability functions of the following random variables:
(a) the number of cabs that are in operative condition;
(b) the number of cabs that are out on a call;
(c) the number of cabs in operative condition that are not out on a call.
(*Hint*: In each case, attach the value of the random variable and the probability to each point of the sample space, and then add the probabilities of the points which correspond to the respective values of the random variable.)

5. Referring to the table on page 169, verify that the values of the probability function for the number of heads in three flips of a balanced coin are given by the formula

$$f(x) = \frac{\binom{3}{x}}{8}$$

for $x = 0, 1, 2,$ and 3.

6. Check whether the following can be looked upon as probability functions (defined in each case only for the given values of x) and explain your answers:

(a) $f(x) = \frac{1}{5}$ for $x = 0, 1, 2, 3, 4$, or 5;

(b) $f(x) = \dfrac{x}{10}$ for $x = 1, 2, 3$, or 4;

(c) $f(x) = \dfrac{x^2}{14}$ for $x = 0, 1, 2$, or 3;

(d) $f(x) = \dfrac{x-3}{3}$ for $x = 1, 2, 3, 4, 5$, or 6.

7. In each case, check whether the given values can be looked upon as the values of the probability function of a random variable which can take on the values 1, 2, 3, 4, and 5, and explain your answers:
(a) $f(1) = 0.15, f(2) = 0.15, f(3) = 0.15, f(4) = 0.15, f(5) = 0.15$;
(b) $g(1) = \frac{1}{9}, g(2) = \frac{2}{9}, g(3) = \frac{3}{9}, g(4) = \frac{2}{9}, g(5) = \frac{1}{9}$;
(c) $p(1) = 0.11, p(2) = 0.28, p(3) = 0.15, p(4) = 0.26, p(5) = 0.20$;
(d) $f(1) = \frac{1}{8}, f(2) = \frac{1}{4}, f(3) = \frac{3}{8}, f(4) = \frac{1}{4}, f(5) = \frac{1}{8}$.

8. Suppose we roll a pair of balanced dice, one green and one red, and we let (1, 3) represent the outcome where the green die comes up 1 and the red die comes up 3, we let (4, 2) represent the outcome where the green die comes up 4 and the red die comes up 2, and so forth.
(a) Draw a figure showing the points of the corresponding sample space, assign the probability $\frac{1}{36}$ to each of its 36 points, and construct a table showing the probabilities of rolling a total of 2, 3, 4, ..., 11, or 12.
(b) Verify that the probabilities of part (a) are also given by the formula

$$f(x) = \frac{6-|x-7|}{36} \quad \text{for } x = 2, 3, \ldots, \text{ or } 12$$

where the **absolute value** $|x-7|$ equals $x-7$ or $7-x$, whichever is positive or zero.

9. Referring to the tables on page 169, find
(a) the mean of the distribution of the number of students who dislike the new record;
(b) the mean of the distribution of the number of students who like the new record minus the number of students who dislike it.
Note that if the result of part (a) is subtracted from the expected number of students who like the new record (which we obtained on page 171), the result is the answer to part (b). What general rule does this suggest?

10. As can easily be verified, the probabilities of getting zero, one, two, three, or four heads in four flips of a balanced coin are, respectively, $\frac{1}{16}, \frac{4}{16}, \frac{6}{16}, \frac{4}{16}$, and $\frac{1}{16}$. Find the mean of this probability distribution.

11. Referring to Exercise 18 on page 148 and its results, find the mean of the probability distribution of the number of opinions the editor will get which favor the publication.

12. Find the means of the probability distributions of the three parts of Exercise 2.

13. Find the means of the probability distributions of the three parts of Exercise 3. What general rule is suggested by the fact that the sum of the results of parts (a) and (b) equals the result of part (c)?

14. Find the means of the probability distributions of the three parts of Exercise 4. What general rule is suggested by the fact that the difference between the results of parts (a) and (b) equals the result of part (c)? See also Exercise 9.

THE BINOMIAL DISTRIBUTION

To consider an example which is similar but slightly more complicated than the coin-tossing example on page 169, suppose that a new car dealer knows from experience that 80 percent of all the new cars he sells require some adjustments or repairs within the first 90 days. Given this information, let us determine the probability function for the number of cars (among three new cars which he sells) that require adjustments or repairs within the first 90 days. If we let R and N denote cars which, respectively, do and do not require adjustments or repairs within the first 90 days, the eight possibilities we shall have to consider are

RRR RRN RNR NRR RNN NRN NNR NNN

These are like the eight possibilities on page 169 with R and N taking the place of H and T, but in contrast to the coin-tossing example, these possibilities are not all equally likely. Since the probability that one of these cars will require adjustments or repairs within the first 90 days is 0.80 or $\frac{4}{5}$, the probability that one of these cars will *not* require adjustments or repairs within the first 90 days is $1-\frac{4}{5}=\frac{1}{5}$, so that the probabilities associated with RRR and RRN, for example, are, respectively, $\frac{4}{5}\cdot\frac{4}{5}\cdot\frac{4}{5}=\frac{64}{125}$ and $\frac{4}{5}\cdot\frac{4}{5}\cdot\frac{1}{5}=\frac{16}{125}$. Similarly, the probabilities associated with NNR and NNN are, respectively, $\frac{1}{5}\cdot\frac{1}{5}\cdot\frac{4}{5}=\frac{4}{125}$ and $\frac{1}{5}\cdot\frac{1}{5}\cdot\frac{1}{5}=\frac{1}{125}$. Thus, the probability that *none* of the cars will require adjustments or repairs within the first 90 days is $\frac{1}{125}$, and the probability that *all three* will require adjustments or repairs within the first 90 days is $\frac{64}{125}$. The probability that *exactly one* of the cars will require adjustments or repairs within the first 90 days is the *sum* of the probabilities associated with RNN, NRN, and NNR, and since each of these is the product of one factor $\frac{4}{5}$ and two factors $\frac{1}{5}$ and, hence, equal to $\frac{4}{125}$, the answer is $3 \cdot \frac{4}{125} = \frac{12}{125}$. Similarly, the probability that *exactly two* of the cars will require adjustments or repairs within the first 90 days is the *sum* of the probabilities associated with RRN, RNR, and NRR, and since each of these is the product of two factors $\frac{4}{5}$ and one factor $\frac{1}{5}$ and, hence, equal to $\frac{16}{125}$, the answer is $3 \cdot \frac{16}{125} = \frac{48}{125}$. All these results are summarized in the following table:

Number of cars requiring adjustments or repairs within first 90 days	*Probability*
0	$\frac{1}{125}$
1	$\frac{12}{125}$
2	$\frac{48}{125}$
3	$\frac{64}{125}$

Note that the four probabilities add up to 1, as they should.

The main thing which this example and the coin-tossing example on page 169 have in common is that in both cases we are interested in the number of "successes" in three "trials." In the coin-tossing example the "trials" were the successive flips of the coin and the "successes" were the ones in which the coin comes up heads; in the new-car example the "trials" were the different cars and the "successes" were the ones which require adjustments or repairs within the first 90 days. As we have already pointed out, this somewhat strange terminology is a carry-over from the days when probability theory was applied only to games of chance; therefore, we shall not let it bother us if a car requiring repairs is referred to as a "success."

There are many applied problems in which we are concerned with the probability of getting x "successes" in n "trials," namely, the probability that an event will take place x times out of n. For instance, we may be interested in the probability that 33 of 60 families (interviewed as part of a large-scale survey) own a color television set, the probability that three of eight laboratory mice will react to a substance injected under their skin, the probability that 183 of 300 adults (interviewed by the Gallup Poll) favor stricter antipollution controls, the probability that 67 of 150 housewives (interviewed in a supermarket) will say that they would not shop there unless they got trading stamps, and so on.

In the coin-tossing and new-car-needing-repairs examples we assumed (without actually saying it in so many words) that *the trials were all independent*, and that *the probability of a "success" remained the same for each trial.* These assumptions could also be justified in the examples of the preceding paragraph if the families, mice, adults, and housewives are suitably selected, but they would not apply to the example on page 144 if we wanted to determine, say, the probability that it will rain on three of five December days which a person spends in the given city. In that case the trials would not be independent. To give an example where the probability of a "success" does not remain constant from trial to trial, we could change the last example and have a person visit the given city for one day each in December, March, May, August, and October.

Whenever the assumptions of independent trials and a constant probability from trial to trial can be applied, the probability function for the number of "successes" is the **binomial probability function**, which we usually refer to simply as the **binomial distribution**. If p denotes the probability of a success on any given trial, the values of this probability function, namely, the probabilities of getting "x successes in n trials," are given by

$$f(x) = \binom{n}{x} \cdot p^x (1-p)^{n-x} \quad \text{for } x = 0, 1, 2, \ldots, \text{ or } n$$

where the $\binom{n}{x}$ are binomial coefficients, which can be calculated according to the formula on page 23, or be looked up in Table II. The reason why we refer to this probability function as "binomial" is that for $x = 0, 1, 2, \ldots,$ and n its values are the successive terms of the binomial expansion of $[p+(1-p)]^n$; this follows immediately from our discussion of the binomial theorem in Chapter 1. Incidentally, since $[p+(1-p)]^n = 1^n = 1$, this also verifies that the sum of all the values of a binomial probability function equals 1, as it should.

To derive the formula for the binomial probability function, we have only to observe that the probability of getting x "successes" and $n-x$ "failures" *in a specific order* is $p^x(1-p)^{n-x}$—there is one factor p for each "success," one factor $1-p$ for each "failure," and these x factors p and $n-x$ factors $1-p$ are all multiplied together in accordance with the generalized multiplication rule for independent events which we gave on page 142. Since this probability applies to any point of the sample space representing x "successes" and $n-x$ "failures" (as we already saw in the new-cars-needing-repairs example on page 176), the desired probability for x "successes" and $n-x$ "failures" *in any order* is simply the product of $\binom{n}{x}$, the number of ways in which we can choose the x trials on which there is to be a "success," and $p^x(1-p)^{n-x}$. This completes the proof.

To illustrate the use of the formula for the binomial distribution, suppose that a chief of police claims that the odds are 4 to 1 that a car stolen in his town will be recovered. Since matters like this are, of course, of interest to insurance companies, suppose that an actuary wants to determine the probability that four of six cars which were stolen in this city will be recovered. Substituting $x = 4$, $n = 6$, and $p = 0.80$ into the formula for the binomial distribution, together with $\binom{6}{4} = 15$ which we obtained from Table II, we get

$$\begin{aligned} f(4) &= \binom{6}{4}(0.80)^4(1-0.80)^{6-4} \\ &= 15(0.80)^4(0.20)^2 \\ &= 0.24576 \end{aligned}$$

or approximately 0.25. Similarly, to find the probability that five of seven reviewers will like a new movie, when the probability that any one of them will like it is 0.60, we substitute $x = 5$, $n = 7$, $p = 0.60$, and $\binom{7}{5} = \frac{7 \cdot 6}{2} = 21$ into the formula, and we get

$$\begin{aligned} f(5) &= \binom{7}{5}(0.60)^5(1-0.60)^{7-5} \\ &= 21(0.60)^5(0.40)^2 \\ &= 0.2613 \end{aligned}$$

rounded to four decimals.

To give an example where we calculate *all* the values of a binomial distribution, suppose that a post office official claims that the ZIP code is missing on one letter in 10. Assuming that this figure is correct, let us find the probabilities that among five letters randomly selected from a big pile the ZIP code will be missing on zero, one, two, three, four, or five. Substituting $n = 5$ and $p = 0.10$ into the formula and, respectively, $x = 0$, 1, 2, 3, 4, and 5, we get

$$f(0) = \binom{5}{0}(0.10)^0(1-0.10)^{5-0} = 0.5905$$

$$f(1) = \binom{5}{1}(0.10)^1(1-0.10)^{5-1} = 0.3280$$

$$f(2) = \binom{5}{2}(0.10)^2(1-0.10)^{5-2} = 0.0729$$

$$f(3) = \binom{5}{3}(0.10)^3(1-0.10)^{5-3} = 0.0081$$

$$f(4) = \binom{5}{4}(0.10)^4(1-0.10)^{5-4} = 0.0004$$

$$f(5) = \binom{5}{5}(0.10)^5(1-0.10)^{5-5} = 0.0000$$

where the results have been rounded to four decimals and the last value is actually 0.00001 and not 0. Note that the odds are almost 3 to 2 that none of the letters will be without a ZIP code. Graphically, this probability distribution is shown in Figure 7.2 in the form of a **histogram**, which

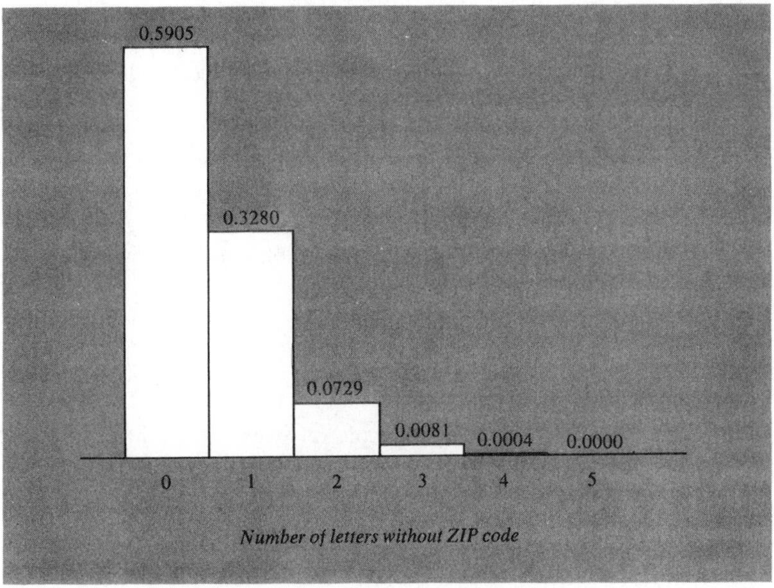

FIGURE 7.2 Histogram of binomial distribution.

gives a good picture of how the total probability of 1 is distributed among the six values which the random variable can take on. The diagram is drawn so that the heights of the rectangles (and, hence, their areas) are equal to the corresponding probabilities.

To carry this example one step further, let us find the mean of the distribution, which will tell us how many letters out of five we can *expect* to be without a ZIP code. Substituting the probabilities together with the corresponding values of the random variable into the formula for μ on page 172, we get

$$\begin{aligned}\mu &= 0(0.5905)+1(0.3280)+2(0.0729)+3(0.0081)+4(0.0004)+5(0.0000)\\ &= 0.4997\end{aligned}$$

or approximately 0.5. This should not come as a surprise—we started with the assumption that 10 percent of all letters lack ZIP codes, and 10 percent of 5 is 0.5. Indeed, the same sort of argument leads to the result that *in general* the mean of the binomial distribution is given by the special formula

$$\mu = n \cdot p$$

In words, *the mean of a binomial distribution is given by the product of the number of trials and the probability of success for each trial.* Thus, if $n = 200$ and $p = 0.25$, we can *expect* 25 percent of 200, or $200(0.25) = 50$ successes, and if $n = 400$ and $p = 0.80$, we can *expect* 80 percent of 400, or $400(0.80) = 320$ successes.

Using the formula $\mu = n \cdot p$ we can also verify the result obtained on page 172 for the mean of the distribution of the number of heads in three flips of a balanced coin. Substituting $n = 3$ and $p = \frac{1}{2}$, we get $\mu = 3 \cdot \frac{1}{2} = \frac{3}{2}$ or $1\frac{1}{2}$, and this agrees with the result which we obtained on page 172.

BINOMIAL TABLES

Suppose that in the car-theft example on page 178 we had wanted to determine the probability that *at least* 12 of 15 cars stolen in that city will be recovered, while still assuming that the probability is 0.80 that a car stolen in that city will be recovered. The answer to this question would be given by the *sum* of the binomial probabilities

$$f(12) = \binom{15}{12}(0.80)^{12}(1-0.80)^{15-12}$$

$$f(13) = \binom{15}{13}(0.80)^{13}(1-0.80)^{15-13}$$

$$f(14) = \binom{15}{14}(0.80)^{14}(1-0.80)^{15-14}$$

and

$$f(15) = \binom{15}{15}(0.80)^{15}(1-0.80)^{15-15}$$

and it should be apparent that the necessary calculations would involve a prohibitive amount of work. The same is true when n is large or we have to find the sum of quite a number of terms—it would be too involved, for example, to use the formula to calculate the probability that four of 8,000 patients will have an undesirable side effect from a new drug, or the probability that fewer than 235 of 760 freshmen entering a given college will manage to graduate.

There are several ways in which binomial probabilities can be *approximated,* but we shall not be able to study them at the level of this book. Unless n is very large, however, there is no need for approximations, since binomial probabilities have been tabulated extensively for various values of n and p. For $n = 2$ to $n = 49$, values of binomial probabilities may be found in a table published by the National Bureau of Standards in 1950, and for $n = 50$ to $n = 100$, they may be found in the H. G. Romig table published by John Wiley & Sons, Inc., in 1953. The table at the end of this book is based on the National Bureau of Standards table, but it is limited to the binomial probabilities for $n = 2$ to $n = 15$, and $p = 0.05$, 0.1, 0.2, 0.3, 0.4, 0.5, 0.6, 0.7, 0.8, 0.9, and 0.95, all rounded to three decimals. As is indicated on page 228, all values omitted in this table are 0.0005 or less.

By using this table, we can immediately answer the question raised at the beginning of this section. For $n = 15$ and $p = 0.80$, the probabilities corresponding to $x = 12$, 13, 14, and 15 are, respectively, 0.250, 0.231, 0.132, and 0.035, and we find that the probability that at least 12 of the 15 stolen cars will be recovered is $0.250 + 0.231 + 0.132 + 0.035 = 0.648$. Also, for the two examples on page 178 where we had $x = 4$, $n = 6$, $p = 0.80$, and $x = 5$, $n = 7$, $p = 0.60$, respectively, we find that the tabular values are 0.246 and 0.261, and this agrees with the earlier results rounded to three decimals.

To consider another problem where the binomial table can be used to advantage, suppose that a distributor of processed foods claims that 70 percent of all housewives are willing to pay 16 cents more for an 8-ounce package of pitted prunes than for an 8-ounce package of unpitted prunes. To test this claim, namely, the *hypothesis* that the number of women in a sample willing to pay the higher price is a random variable having the binomial distribution with $p = 0.70$, we would ordinarily take a fairly large sample. Since our table goes only to $n = 15$, however,, let us use this value for the number of housewives who are to be interviewed, and base our decision more or less arbitrarily on the following criterion:

Reject the distributor's claim if fewer than seven or more than 13 of the 15 women say that they are willing to pay the higher price for the pitted prunes; otherwise, accept it.

Like all decision criteria which are based on sample data, this criterion is not infallible. For instance, it is possible to get a sample of 15 women including fewer than seven or more than 13 who are willing to pay the higher price *even though the distributor's claim is true,* that is, even though $p = 0.70$. In fact, it can be seen from Table III that the binomial probabilities for $n = 15$, $p = 0.70$, and $x = 0, 1, 2, 3, 4, 5, 6, 14$, and 15 are, respectively, 0.000, 0.000, 0.000, 0.000, 0.001, 0.003, 0.012, 0.031, and 0.005, so that the probability of *erroneously rejecting the distributor's claim* is $0.001+0.003+0.012+0.031+0.005 = 0.052$.

A BAYESIAN APPLICATION

There are many situations in which we must modify probability judgments in the light of new evidence, that is, after obtaining additional or different kinds of information. Suppose, for example, that two business partners, Mr. Allen and Mr. Bates, are planning to form a company to build and operate a chain of drive-in restaurants. While talking to their banker to arrange for the necessary financing, Mr. Allen claims that the odds are 4 to 1 that any person they hire to operate one of these restaurants will stay with them for at least a year, while Mr. Bates feels that the odds should be only 3 to 2. Suppose also that on the basis of his past dealings with these two businessmen, the banker feels that in matters like this Mr. Allen is three times as reliable as Mr. Bates. What we would like to know is *how the banker's opinion about the reliability of Mr. Allen and Mr. Bates would be affected if it turned out that among 10 persons they hire to operate their restaurants only one fails to stay with them at least a year.*

If we let B denote the event that Mr. Allen is right and B' the event that Mr. Bates is right, which assumes that one or the other must be right, we find that the *prior probabilities* which the banker assigns to these two events are $P(B) = \frac{3}{4}$ and $P(B') = \frac{1}{4}$, for this makes Mr. Allen "three times as reliable" as Mr. Bates. Then, if A is the event that among 10 persons they hire to operate their restaurants nine stay with them for at least a year, the probabilities we shall want to determine are the *posterior probabilities* $P(B \mid A)$ and $P(B' \mid A)$. They express the banker's feelings about the reliability of Mr. Allen and Mr. Bates *after he has the additional information about the occurrence of event A.* To obtain these probabilities we must use Bayes' rule, and thus we shall first have to calculate the two probabilities

$P(A \mid B)$ and $P(A \mid B')$. Since $P(A \mid B)$ is the probability of "nine successes in 10 trials" when Mr. Allen is right, namely, when $p = \frac{4}{4+1} = 0.80$, we find that

$$P(A \mid B) = \binom{10}{9}(0.80)^9(1-0.80)^{10-9} = 0.268$$

and since $P(A \mid B')$ is the probability of "nine successes in 10 trials" when Mr. Bates is right, namely, when $p = \frac{3}{3+2} = 0.60$, we find that

$$P(A \mid B') = \binom{10}{9}(0.60)^9(1-0.60)^{10-9} = 0.040$$

Then, if we substitute these values together with $P(B) = \frac{3}{4}$ and $P(B') = \frac{1}{4}$ into the formula for Bayes' rule as given on page 160, we get

$$P(B \mid A) = \frac{\frac{3}{4}(0.268)}{\frac{3}{4}(0.268)+\frac{1}{4}(0.040)} = 0.95$$

and it follows that $P(B' \mid A) = 1-0.95 = 0.05$. It is interesting to note how the additional evidence about the 10 persons hired to operate the restaurants has affected the banker's opinion about the reliability of Mr. Allen and Mr. Bates. Whereas at first he would have given odds of 3 to 1 that Mr. Allen is right and Mr. Bates is wrong, the weight of the direct information has "lengthened" these odds to 95 to 5, or 19 to 1.

EXERCISES

1. Use the formula for the binomial distribution to find
   (a) the probability of getting exactly two 3's in five rolls of a balanced die;
   (b) the probability of getting at most two 3's in five rolls of a balanced die;
   (c) the probability of getting exactly six heads in eight flips of a balanced coin;
   (d) the probability of getting at least six heads in eight flips of a balanced coin.

2. A traffic engineer claims that two in 10 automobile accidents are due to driver fatigue. If this is true, find the probability that among five accidents exactly one is due to driver fatigue

(a) by using the formula for the binomial distribution;
(b) referring to Table III.

3. A multiple-choice test consists of six questions, with each having four possible answers of which only one is correct. If a student answers each question by cutting a deck of cards and checking the first, second, third, or fourth alternative depending on whether he gets a spade, heart, diamond, or club, find the probabilities that he will get
(a) nothing but wrong answers;
(b) one right answer and five wrong answers;
(c) two right answers and four wrong answers.
Also use the results of parts (a), (b), and (c) to find the probability that he will get
(d) at least three right answers.

4. In a given county, 40 percent of all losses due to fraudulent, dishonest, or criminal acts are indemnified by insurance. If eight such cases are randomly selected from court files, find the probability that *less than* three of them were indemnified by insurance
(a) by using the formula for the binomial distribution;
(b) referring to Table III.

5. In a Western city, incompatibility is given as the legal reason for 70 percent of all divorce cases. Find the probability that four of the next six cases filed in this city will claim to be due to incompatibility
(a) by using the formula for the binomial distribution;
(b) referring to Table II.

6. It is expected that 30 percent of the mice used in an experiment will become very active within a minute after having been administered an experimental drug. Should the person who conducts this experiment be greatly surprised if he administers the drug to eight mice and none of them become very active within a minute? Calculate the binomial probability that this will happen *by chance,* that is, even though the expected percentage is correct.

7. The owner of a small motel with eight units has five color television sets which he installs upon request at an extra charge. If there is a fifty-fifty chance that any one of his guests wants a color television set, use the formula for the binomial distribution to find the probability that on a night when all the units are occupied, there will be more requests for sets than there are sets.

8. Suppose that the odds are 3 to 2 that a housewife who lets a cosmetics salesperson into her house will end up making a purchase.

Find the probability that *at least* five of seven housewives who let such a salesperson into their home will end up making a purchase
(a) by using the formula for the binomial distribution;
(b) referring to Table III.

9. It is clained that among all drivers whose cars are equipped with seatbelts, only 75 percent use them on long trips. Use the formula for the binomial distribution to find the probabilities that in four cars (equipped with seatbelts) which pass through a toll station of a parkway
(a) none of the drivers are using their seatbelts;
(b) all of the drivers are using their seatbelts;
(c) only one of the drivers is using his seatbelt.
Also use the results of parts (a), (b), and (c) to find the probability that
(d) at least two but not all of the drivers are using their seatbelts.

10. Use Table III to verify the values of the binomial distribution with $n = 5$ and $p = 0.10$ given on pages 178 and 179 rounded to three decimals.

11. A social scientist claims that only 40 percent of those who are capable of college work go to college. If this is so, use Table III to find the probabilities that of 12 high schools seniors capable of college work
(a) at least half will go to college;
(b) not more-than four will go to college;
(c) anywhere from three through eight, inclusive, will go to college.

12. According to a survey, 60 percent of all families with incomes of over $12,000 a year have at least two television sets. Use Table III to find the probabilities that among 14 such families
(a) all will have at least two television sets;
(b) at most five will have at least two television sets;
(c) at least seven will have at least two television sets;
(d) anywhere from five through 11, inclusive, will have at least two television sets;
(e) fewer than four or more than 12 will have at least two television sets.

13. Suppose that a civil service examination is designed so that 80 percent of all persons with an I.Q. over 90 can pass it. Use Table III to find the probabilities that among 15 persons with an I.Q. over 90 who take the test
(a) fewer than 10 will pass;
(b) none will pass;
(c) at least nine will pass;
(d) anywhere from nine through 12, inclusive, will pass.

14. Suppose that a political scientist wants to determine what proportion of all registered voters favor a tax on foreign imports. He decides to interview a sample of 10 voters, chosen at random and then claim that *the true proportion is greater than the proportion he obtains in the sample minus 0.10*. For instance, if his sample includes six voters who favor a tax on foreign imports, he will claim that the true proportion is greater than $\frac{6}{10} - 0.10 = 0.50$. Use Table III to find the probabilities that the claim he will make will be *wrong* when

(a) the true proportion is 0.70;

(b) the true proportion is 0.40.

(*Hint*: for each number of "successes" check whether the corresponding sample proportion *minus 0.10* is actually exceeded by the true proportion.)

15. Suppose that the owner of a large nursery wants to estimate what proportion of certain palm trees will be able to stand temperatures down to 10 degrees. He has 13 of these palm trees which he exposes to temperatures down to 10 degrees, and if x of them show no damage he will claim that the interval from $\frac{x}{15}$ to $\frac{x+2}{15}$, inclusive, contains the *true proportion* (of such palms which can stand temperatures down to 10 degrees). For instance, if 10 of the trees show no damage, he will claim that the interval from $\frac{10}{15} = \frac{2}{3}$ to $\frac{10+2}{15} = \frac{4}{5}$ contains the true proportion. Use Table III to find the probabilities that the claim he will make will be *wrong* when

(a) the true proportion is 0.20;

(b) the true proportion is 0.30.

(*Hint*: Calculate the interval for each value of x and check whether it contains the true proportion.)

16. Verify for the example on page 182 that the probability of erroneously accepting the distributor's claim when only 20 per cent of all housewives are willing to pay the higher price is 0.018.

17. A quality control inspector wants to check whether (in accordance with specifications) 95 percent of the electronic components shipped by his company are in good working condition. To this end, he randomly selects 12 from each large lot ready to be shipped and passes it only if they are *all* in good working condition; otherwise, each of the components in the lot is checked.

(a) What is the probability that he will commit the error of holding a lot for further inspection even though 95 percent of the components are in good working condition?

(b) What is the probability that he will commit the error of letting a lot pass through even though only 80 percent of the components are in good working condition?

18. A food distributor claims that 80 percent of his 6-ounce cans of mixed nuts contain at least three pecans. To check on this, a consumer testing service decides to take 10 of these 6-ounce cans of mixed nuts from a very large production lot, and reject the claim if fewer than seven of them contain at least three pecans.

(a) What is the probability that the testing service will commit the error of rejecting the claim even though it is true?

(b) What is the probability that the testing service will commit the error of accepting the claim when in reality only 50 percent of the cans contain at least three pecans?

(c) What is the probability that the testing service will commit the error of accepting the claim when in reality only 30 percent of the cans contain at least three pecans?

19. Suppose that a faculty committee wants to investigate the claim that 60 percent of the students attending a very large university are opposed to the administration proposal to increase their fees to raise funds for a new stadium. They decide to interview 15 students, chosen at random, and accept the claim if and only if anywhere from six through 11, inclusive, are opposed to the plan.

(a) What is the probability that they will commit the error of rejecting the claim even though it is true?

(b) What is the probability that they will commit the error of accepting the claim when in reality only 50 percent of the students are opposed to the plan?

(c) What is the probability that they will commit the error of accepting the claim when in reality 80 percent of the students are opposed to the plan?

20. The manager of a restaurant figures that 40 percent of his customers order fried chicken. Referring to Table III, find the probabilities that 0, 1, 2, 3, ..., or 14 customers out of 14 will order fried chicken, and use them to determine how many of the 14 customers the manager can *expect* to order fried chicken. Check the result with the formula $\mu = n \cdot p$ and also draw a histogram which pictures this probability distribution.

21. A real estate salesman figures that 90 percent of all the women to whom he shows a certain model home will complain about the size of the kitchen. Referring to Table III, find the probabilities that 0, 1, 2, ..., or 12 women out of 12 to whom he shows the model home will complain about the size of the kitchen and draw a histogram to picture this distribution. Also use these probabilities to determine how many of the 12 women the salesman can *expect* to complain about the size of the kitchen, and verify the result by means of the formula $\mu = n \cdot p$.

22. Find the mean of the binomial distributions of the following random variables:

(a) the number of *heads* obtained in 436 flips of a balanced coin;
(b) the number of 6's in 45 rolls of a balanced die;
(c) the number of persons in a sample of 400 who remember what products were advertised on a certain television program, if the probability that any one of them will remember the products is 0.32;
(d) the number of seeds that will germinate in a package of 120, if the probability that any one of them will germinate is 0.85;
(e) the number of families that visit Disneyland among 250 families with children vacationing in Southern California, if the probability that any such family will visit Disneyland is 0.84;
(f) the number of drivers among 250 stopped at a roadblock who have not received a traffic ticket within the last 12 months, if the probability that any one of them has not received a traffic ticket within the last 12 months is 0.64.

23. The mean of a probability distribution not only tells us what to *expect*, or what we will get *on the average*, but in most cases we can be pretty sure that we will get a value that is *close to the mean*.

(a) Use Table III to find the probability that a value of a random variable having the binomial distribution with $n = 15$ and $p = 0.50$ will differ from the mean by not more than 1.5.
(b) Use Table III to find the probability that a value of a random variable having the binomial distribution with $n = 12$ and $p = 0.40$ will differ from the mean by less than 2.0.
(c) Use Table III to find the probability that a value of a random variable having the binomial distribution with $n = 14$ and $p = 0.20$ will differ from the mean by more than 1.8.

24. Mr. Butler and Mr. Daniels intend to go pheasant hunting, using Mr. Butler's dog. According to Mr. Butler the probability that his dog will find a downed bird is 0.80 and according to Mr. Daniels this probability is only 0.60. In the absence of any further information we might judge that, since it is Mr. Butler's dog, he is four times as likely to be right as Mr. Daniels. Assuming that one or the other must be right, how would this judgment be affected if on the hunting trip the dog actually finds only four of the nine pheasants they downed. (*Hint*: Use Bayes' rule and look up the required binomial probabilities in Table III.)

25. Mr. Ames, Mr. Brown, and Mr. Charles are planning to open a chain of donut shops. According to Mr. Ames the probability that any one of them will show a profit during the first year is 0.10, and according to Mr. Brown and Mr. Charles this probability is 0.05 and 0.20, respectively. An impartial expert feels that Mr. Ames is three times as likely to be right as Mr. Brown, and that Mr. Brown is twice as likely to be right as Mr. Charles. If a survey shows that among 12 such donut shops only one showed a profit during the first year, find

(a) the *prior probabilities* which the impartial expert assigns to $p = 0.10$, $p = 0.05$, and $p = 0.20$, assuming that one of them must be right;

(b) the corresponding *posterior probabilities* which we get with the use of Bayes' rule, if the information about the 12 donut shops is taken into consideration. Use Table III to look up the necessary binomial probabilities.

THE HYPERGEOMETRIC DISTRIBUTION

In Chapter 6 we talked about sampling *with and without replacement* to illustrate the multiplication rules for independent and dependent events. When we are interested in the total number of successes, the binomial distribution applies to sampling with replacement, and to introduce a corresponding distribution which applies to sampling without replacement, let us consider the following example: Suppose that certain cameras are shipped from overseas in lots of 24. When they arrive at the importer's warehouse, an inspector randomly selects four cameras from each lot and the whole lot is accepted if all four are in perfect condition: otherwise, each camera is inspected individually (and repaired, if necessary, at a considerable cost). It is easy to see that this kind of **sampling inspection** involves certain risks—it is possible, for example, for a lot to "fail" the preliminary inspection even though only one of the 24 cameras has a slight defect, and it is possible for a lot to "pass" the preliminary inspection even though 20 of the 24 cameras have serious defects. Of course, many questions arise in situations like this, but it will serve our purpose to investigate only the probability that a lot will "pass" the preliminary inspection when, say, three of the 24 cameras are not in perfect condition. This means that we shall have to find the probability of getting four successes (cameras in perfect condition) in four trials, and we might be tempted to argue that since 21 of the 24 cameras are in perfect condition, the probability of getting such a camera is $\frac{21}{24} = \frac{7}{8}$, and hence that the probability of "four successes in four trials" is

$$f(4) = \binom{4}{4}\left(\frac{7}{8}\right)^4\left(1-\frac{7}{8}\right)^{4-4}$$
$$= 1\left(\frac{7}{8}\right)^4$$
$$= 0.586$$

This result may be acceptable in the sense of an *approximation*, but what we have done is not really legitimate. *The trials are not independent and the binomial distribution does not apply.* Although the probability *is* $\frac{21}{24}$ that the first camera which is inspected is in perfect condition, the corresponding probabilities for the second, third, and fourth are only $\frac{20}{23}$, $\frac{19}{22}$, and $\frac{18}{21}$, for after each choice the number of cameras and the number of cameras in perfect condition are reduced by 1. Thus, the correct probability of getting four perfect cameras (when 21 of the 24 cameras are in perfect condition) is

$$\frac{21}{24}\cdot\frac{20}{23}\cdot\frac{19}{22}\cdot\frac{18}{21} = \frac{143{,}640}{255{,}024}$$

or approximately 0.563. The difference between 0.586 and 0.563 may not be very large and it may not be of much practical significance, but this does not make up for the fact that we were *wrong* in using the formula for the binomial distribution. That formula would have been correct if each camera had been replaced before the next one was chosen for inspection, but this is hardly what anyone would do in actual practice.

Appropriate for sampling without replacement is the so-called **hypergeometric distribution**. To introduce the formula for this probability function, let us observe that in the above example we could have argued that four of the 24 cameras can be chosen in $\binom{24}{4} = 10{,}626$ ways, four of the perfect cameras can be chosen in $\binom{21}{4} = 5{,}985$ ways, so that the probability of getting four cameras in perfect condition is

$$\frac{\binom{21}{4}}{\binom{24}{4}} = \frac{5{,}985}{10{,}626}$$

or approximately 0.563 *provided the 10,626 possibilities can be regarded as equally likely.* As should be noted, this result agrees with the one which we obtained before.

In general, if n objects are to be chosen from a set consisting of a objects of one kind (call them "successes") and b objects of another kind (call them "failures"), and we are interested in the probability of getting "x successes and $n-x$ failures," we can similarly argue that the x successes can be chosen in $\binom{a}{x}$ ways, the $n-x$ failures can be chosen in $\binom{b}{n-x}$ ways, and hence the x successes and $n-x$ failures can be chosen in $\binom{a}{x}\cdot\binom{b}{n-x}$ ways. Also, n objects can be chosen from the whole set in $\binom{a+b}{n}$ ways, and if we regard these as equally likely, the probability of getting "x successes and $n-x$ failures" is

$$f(x) = \frac{\binom{a}{x}\cdot\binom{b}{n-x}}{\binom{a+b}{n}} \quad \text{for } x = 0, 1, 2, \ldots, \text{ or } n$$

This is the formula for the *hypergeometric distribution,* and we should add that *it applies only when x does not exceed a and $n-x$ does not exceed b*; clearly, we cannot very well get more successes (or failures) than there are to begin with in the whole set.

To consider another example, suppose that 100 of the 240 inmates of a prison have radical political views. If five of them are randomly chosen to appear before a legislative committee which is investigating prison conditions, what is the probability that *at least* one of them will have radical political views? The probability that at least one of them will have radical views is equal to 1 *minus* the probability that *none* of them has radical political views, namely,

$$1 - \frac{\binom{100}{0}\cdot\binom{140}{5}}{\binom{240}{5}} = 1 - \frac{140\cdot139\cdot138\cdot137\cdot136}{240\cdot239\cdot238\cdot237\cdot236}$$

$$= 1 - 0.0655$$

$$= 0.9345$$

Having used the formula for the hypergeometric distribution with $x = 0$,

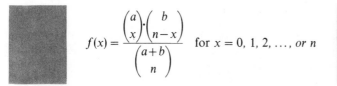

$n = 5$, $a = 100$, and $b = 140$ in this example, let us see how much of an error we would have made if we had erroneously used the formula for the binomial distribution with $n = 5$ and $p = \frac{100}{240} = \frac{5}{12}$. Substituting these values together with $x = 0$ into the formula for the binomial distribution, we get

$$f(0) = \binom{5}{0}\left(\frac{5}{12}\right)^0\left(1-\frac{5}{12}\right)^{5-0}$$
$$= 1\left(\frac{7}{12}\right)^5$$

which is approximately 0.07, and we would thus have arrived at the result that the probability is $1-0.07 = 0.93$ (instead of 0.9345) that at least one of the five prisoners chosen to appear before the committee will hold radical political views. This approximation is quite close, and in actual practice we often use the binomial distribution with $p = \dfrac{a}{a+b}$ to approximate hypergeometric probabilities. Mainly, we do this because the binomial distribution has been tabulated much more extensively. As a general rule of thumb, it is advisable to use this kind of approximation only *when n does not constitute more than 5 percent of $a+b$.*

The mean of a hypergeometric distribution can be calculated like that of any other probability distribution in accordance with the formula on page 172, and we already did so, in fact, in the example pertaining to the mystery books on page 172. When the person grabs two of the 15 mysteries, four of which he has not yet read, the number of books he gets which he has not yet read is a random variable having the hypergeometric distribution with $n = 2$, $a = 4$, and $b = 11$, and as we saw on page 172, the mean of the this distribution is $\mu = \frac{8}{15}$. This result could also have been obtained by making use of the following special formula for *the mean of a hypergeometric distribution*

$$\mu = n \cdot \frac{a}{a+b}$$

We shall not prove this formula, but let us point out that it is like the one for the mean of the binomial distribution with $\dfrac{a}{a+b}$ substituted for p. In any case, if we substitute $n = 2$, $a = 4$, and $b = 11$ into the formula, we get

$$\mu = 2 \cdot \frac{4}{4+11} = 2 \cdot \frac{4}{15} = \frac{8}{15}$$

for the expected number of books which the person has not yet read, and this agrees with the result which we obtained on page 172.

THE GEOMETRIC DISTRIBUTION

In Exercise 35 on page 151 the reader was asked to find the probabilities that a person will get his driver's license on the third try or on the fifth, and this was straightforward, requiring only the multiplication rule for events which may or may not be independent. The situation would be more complicated, however, if we considered as a random variable *the number of the trial* on which a person finally passes the test. Since a person can pass the test on the first try, the second try, the third try, and so on, there are as many possibilities as there are positive integers, and we say that the sample space is **countably infinite**, or that it is infinite yet **discrete**. This is to distinguish it from situations where there is actually a **continuum** of possibilities, say, when the sample space consists of all the points on a line segment or a line, or all the points of a region in the plane. This would be the case, theoretically at least, when we throw a dart at a circular target or, for that matter, whenever we are concerned with quantities that are measured on a continuous scale—lengths, weights, time, speed, temperature, and so forth.

When there is a countable infinity of possibilities, the postulates of probability will require a minor modification. On page 115 we already generalized Postulate 3 so that it applied to k mutually exclusive events, and now we have to reformulate it so that it applies to *any number of mutually exclusive events.* Further complications arise when there is a continuum of possibilities, but this is something we shall not discuss in this book.

In the driver's test example the successive trials were *dependent,* but this would not be the case, for example, if we were interested in the number of the trial on which a balanced coin comes up heads for the first time, the number of cans of mixed nuts we may have to open until we finally find a Brazil nut, or perhaps the number of burglaries a burglar may be able to commit until he gets caught for the first time. When we are thus satisfying the conditions underlying the binomial distribution (that is, independent trials and the same probability p of success for each trial), the probability that *the first success will occur on the xth trial* is given by

$$f(x) = p(1-p)^{x-1} \quad \text{for } x = 1, 2, 3, \ldots$$

This defines the **geometric probability function** which, incidentally, owes its name to the fact that for successive values of x the corresponding probabilities constitute a geometric progression. The formula for the geometric distribution is easy to prove: For the first success to be on the xth trial, we must first have $x-1$ failures, the probability of which is $(1-p)^{x-1}$, and then a success, the probability of which is p. Thus, the desired probability is given by the product of $(1-p)^{x-1}$ and p, namely, $p(1-p)^{x-1}$.

To consider a few examples, let us first calculate the probability that a person gets the *first heads* on the twelfth flip of a balanced coin, and the probability that he gets the *first 6* on the eighth roll of a balanced die. For the *first heads* we substitute $x = 12$ and $p = \frac{1}{2}$, and get

$$\begin{aligned} f(x) &= \tfrac{1}{2}(1-\tfrac{1}{2})^{12-1} \\ &= \tfrac{1}{2}(\tfrac{1}{2})^{11} \\ &= \frac{1}{4{,}096} \end{aligned}$$

or approximately 0.00024, and for the *first 6* we substitute $x = 8$ and $p = \frac{1}{6}$, and get

$$\begin{aligned} f(x) &= \tfrac{1}{6}(1-\tfrac{1}{6})^{8-1} \\ &= \tfrac{1}{6}(\tfrac{5}{6})^{7} \\ &= \frac{78{,}125}{1{,}679{,}616} \end{aligned}$$

or approximately 0.047. Also, if the probability that a burglar will get caught on any given "job" is 0.20, then the probability that he will get caught for the first time on his fourth "job" is

$$\begin{aligned} f(x) &= (0.20)(1-0.20)^{4-1} \\ &= (0.20)(0.80)^{3} \\ &= 0.1024 \end{aligned}$$

When the probability p of a success is *high* (close to 1), it stands to reason that we can *expect* the first success to happen fairly soon; correspondingly, when p is *small* (close to 0), we should not be surprised if we have to wait quite some time. All this is expressed formally by the fact that the *mean of the geometric distribution* is given by the formula

$$\mu = \frac{1}{p}$$

For a balanced coin, for example, it takes *on the average* $\mu = \frac{1}{1/2} = 2$ flips to get the first heads; for a balanced die it takes *on the average* $\mu = \frac{1}{1/6} = 6$ rolls to get the first 6; and the burglar of the above example can *expect* that it will take $\mu = \frac{1}{0.20} = 5$ "jobs" till he will get caught. (Derivations of the formula for the mean of the geometric distribution and the formulas for the means of the binomial and hypergeometric distributions can be found in most standard textbooks on mathematical statistics.)

THE MULTINOMIAL DISTRIBUTION

An important generalization of the binomial distribution arises when there are more than two possible outcomes for each trial, the probabilities of the respective outcomes remain the same for each trial, and the trials are all independent. This is the case, for example, for repeated rolls of a die, where each roll has six possible outcomes and each outcome has the probability $\frac{1}{6}$, and it applies to the situation described in the beginning of Chapter 4 where each student is asked whether he likes a new record, dislikes it, or doesn't care, or when a U.S.D.A. inspector grades beef as "Prime," "Choice," "Good," "Commercial," or "Utility" as in the example on page 116.

To derive a formula which applies to situations like this, suppose there are n independent trials, with each trial permitting k mutually exclusive alternatives whose respective probabilities are $p_1, p_2, \ldots$, and p_k. Since one of the alternatives must occur, $p_1 + p_2 + \ldots + p_k = 1$. The random variables with which we shall be concerned are x_1, the number of times the first alternative occurs (namely, the number of "successes of the first kind"), x_2, the number of times the second alternative occurs (namely, the number of "successes of the second kind"), $\ldots$, and x_k, the number of times the kth alternative occurs (namely, the number of "successes of the kth kind"). Of course, since there are n trials, $x_1 + x_2 + \ldots + x_k = n$. Proceeding as in the derivation of the formula for the binomial distribution, let us observe that the probability of getting x_1 successes of the first kind, x_2 successes of the second kind, $\ldots$, and x_k successes of the kth kind *in a specific order* is

$$p_1^{x_1} \cdot p_2^{x_2} \cdot \ldots \cdot p_k^{x_k}$$

There is one factor p_1 for each success of the first kind, one factor p_2 for each success of the second kind, ..., and these x_1 factors p_1, x_2 factors $p_2, \ldots$, are all multiplied together in accordance with the generalized multiplication rule for independent events which we gave on page 142.

Since this probability applies to any point of the sample space representing x_1 successes of the first kind, x_2 successes of the second kind, and so on, the desired probability for that many successes of each kind *in any order* is simply the product of $p_1^{x_1} \cdot p_2^{x_2} \cdot \ldots \cdot p_k^{x_k}$ and the total number of ways in which that many successes of each kind be obtained in n trials. This takes us back to the section of Chapter 1, where we began with rearranging the letters in the word "room." As we saw on page 19, n objects of which x_1 are alike, x_2 others are alike, ..., and x_k othere are alike, can be arranged in

$$\frac{n!}{x_1! \cdot x_2! \cdot \ldots \cdot x_k!}$$

different permutations, and this is precisely the number of ways in which x_1 successes of the first kind, x_2 successes of the second kind, ..., and x_k successes of the kth kind can be obtained in n trials. Thus, the probability of getting x_1 successes of the first kind, x_2 successes of the second kind, ..., and x_k successes of the kth kind is given by the product

$$\frac{n!}{x_1! \cdot x_2! \cdot \ldots \cdot x_k!} \cdot p_1^{x_1} \cdot p_2^{x_2} \cdot \ldots \cdot p_k^{x_k}$$

which defines the so-called **multinomial distribution.**

To give an example, suppose that the manager of a large television station tells a prospective advertiser that on Saturday nights his station, Station X, has 60 percent of the viewing audience, while his two competitors, Stations Y and Z, have, respectively, 30 percent and 10 percent. If this is so, what is the probability that among nine families watching television on a Saturday evening, five will be watching Station X, two will be watching Station Y, and two will be watching Station Z? Thus, $n = 9$, $x_1 = 5$, $x_2 = 2$, $x_3 = 2$, $p_1 = 0.60$, $p_2 = 0.30$, and $p_3 = 0.10$, and substitution into the formula yields

$$\frac{9!}{5! \cdot 2! \cdot 2!} \cdot (0.60)^5 (0.30)^2 (0.10)^2 = 0.05$$

(This really does not prove anything with regard to the station manager's claim, for there are many possibilities, and none of them has a very high probability.)

To consider another example, suppose that a supermarket carries four grades of ground beef, and that the probabilities that a housewife will choose the poorest, third best, second best, or best are, respectively, 0.10, 0.25, 0.50, and 0.15. To find the probability that among 12 randomly chosen housewives buying ground beef at this market one will choose the poorest kind, two will choose the third best kind, seven will buy the second best kind, and two will choose the best kind, we have only to substitute $n = 12$, $x_1 = 1$, $x_2 = 2$, $x_3 = 7$, $x_4 = 2$, $p_1 = 0.10$, $p_2 = 0.25$, $p_3 = 0.50$, and $p_4 = 0.15$ into the formula, and we get

$$\frac{12!}{1!\cdot 2!\cdot 7!\cdot 2!}\cdot(0.10)^1(0.25)^2(0.50)^7(0.15)^2 = 0.026$$

EXERCISES

1. Find the probability that an I.R.S. auditor will catch two income tax returns with illegitimate deductions, if he randomly selects six returns from among 20 returns of which eight contain illegitimate deductions.

2. A quality control engineer inspects a random sample of three toasters from each lot of 24 toasters that is ready to be shipped. If such a lot contains six toasters with slight defects, find the probabilities that the inspector's sample will
(a) contain only perfect toasters;
(b) contain one defective toaster;
(c) contain two defective toasters;
(d) contain three defective toasters.
Also draw a histogram of this hypergeometric distribution.

3. Use the probabilities obtained in Exercise 2 to determine the mean of this hypergeometric distribution, and check the result with the special formula given on page 192.

4. When she buys a dozen eggs, Mrs. Jones always randomly selects and inspects two of the eggs for cracks, and she looks for another carton if at least one of the eggs has a crack. What are the probabilities that she will buy
(a) a carton with one cracked egg;
(b) a carton with two cracked eggs;
(c) a carton with five cracked eggs?
Also find the probabilities that she will *not* buy
(d) a carton with one cracked egg;
(e) a carton with three cracked eggs.

5. A secretary is supposed to send four of 10 letters by air mail. If she gets them all mixed up and randomly puts air mail stamps on four of them, what are the probabilities that
(a) none of the letters which should go by air mail gets an air mail stamp;
(b) one of the letters which should go by air mail gets an air mail stamp;
(c) two of the letters which should go by air mail get an air mail stamp;
(d) three of the letters which should go by air mail get an air mail stamp;
(e) all four of the letters which should go by air mail get an air mail stamp?
Also draw a histogram of this hypergeometric distribution.

6. Use the probabilities obtained in Exercise 5 to determine the mean of this hypergeometric distribution, and check the result by means of the special formula on page 192.

7. A collection of 15 Spanish gold doubloons contains four counterfeits. If two of these coins are randomly selected to be sold at auction, what are the probabilities that
(a) neither coin is a counterfeit;
(b) one of the two coins is a counterfeit;
(c) both coins are counterfeits?

8. Use the probabilities obtained in Exercise 7 to determine the mean of this hypergeometric distribution, and check the result by means of the special formula on page 192.

9. With reference to the illustration on page 189, find the probabilities that
(a) a lot will "pass" the preliminary inspection even though six of the 24 cameras are not in perfect condition;
(b) a lot will "fail" the preliminary inspection when only one of the 24 cameras is not in perfect condition.

10. Among the 80 cities which a political organization is considering for its next four annual conventions, 32 are in the Western part of the United States. If, to avoid arguments, the selection is left to chance (that is, random sampling without replacement), what is the probability that none of the four conventions will be held in the Western part of the United States? How big an error would we make if we approximated this probability with the binomial probability of "zero successes in four trials" when $p = \frac{32}{80} = 0.40$?

11. Among the 120 employees of a company 80 are union members while the others are not. If five of the employees are to be chosen by lot to serve on a committee which administrates the pension fund, what is the probability that three of them will be union members while the other two are not? How big an error would we make if we approximated this probability with the binomial probability of "three successes in five trials" when $p = \frac{80}{120} = \frac{2}{3}$?

12. In a file of 400 invoices exactly eight contain errors. If an auditor randomly selects three invoices from this file, what is the probability that none of them will contain an error? How big an error would we make if we approximated this probability with the binomial probability of "zero successes in three trials" when $p = \frac{8}{400} = 0.02$?

13. Suppose the probability is 0.10 that any person will believe a rumor about the private life of a certain politician. What is the probability that the sixth person to hear this rumor will be the first one to believe it?

14. In a "torture test," a watch is dropped from a tall building until it breaks. If the probability that it will break any time that it is dropped from the building is 0.12 (and we assume independence), what is the probability that the watch will finally break the fourth time it is dropped?

15. In the filming of a television commercial, the probability that a given actress will get her lines straight on any one take is 0.60. Assuming independence, what is the probability that she will finally get her lines straight on the fifth take?

16. An expert rifleman hits a moving target 99 percent of the time. What is the probability that he will miss the target for the first time on his fourth shot?

17. A life insurance salesman regularly calls on families that have recently moved to the San Diego area, even though the odds are 9 to 1 against his making a sale.
   (a) What is the probability that he will make his first sale of the week to the fourth family on which he calls?
   (b) What is the probability that he will make his first sale to the seventh family on which he calls?
   (c) What is the probability that he will still be waiting to make his first sale of the week after having visited 10 families?

18. Find the probabilities that the *first heads* comes on the first, second, third, . . ., or tenth flip of a balanced coin. Assuming that for later trials the probabilities are so small that they can be ignored, calculate the mean of this geometric distribution and compare the result with the actual value as given by the special formula on page 195. Also draw a histogram of this distribution.

19. The same sort of reasoning which led to the formula for the geometric distribution can also be used to handle problems in which we are concerned with the rth success occurring on the xth trial. We multiply the binomial probability of $r-1$ successes on the first $x-1$ trials by the probability of a success on the xth trial. Use this method in connection with Exercise 13 to find the probability that the fifteenth person to hear the rumor will be the fourth to believe it. (*Hint*: refer to Table III.)

20. Referring to the example on page 194 and the method suggested in the preceding exercise, find the probability that the second time the burglar gets caught is on his tenth job. (*Hint*: refer to Table III.)

21. Referring to Exercise 17 and the method suggested in Exercise 19, find the probability that the insurance salesman will make his second sale to the fourteenth family he visits. (*Hint*: refer to Table III.)

22. The same sort of reasoning which led to the formula for the geometric distribution can also be used to handle problems in which we are concerned with the first success occurring on the xth trial, but sampling is *without replacement*. We multiply the probability of zero successes on the first $x-1$ trials by the probability of a success on the xth trial, and the only difference is that the first of these probabilities is given by the formula for the hypergeometric distribution instead of the one for the binomial distribution. Use this method in the following problem: A person has 12 keys, of which two will open his front door. If he tries them in a random order but not the same key more than once, what is the probability that the fifth key he tries will be the first one to open his front door?

23. With reference to Exercise 22 on page 165, suppose that the practical joker has put mustard in two of the six "jelly" donuts. Use the method suggested in the preceding exercise to find the probability that the third person who takes one of these donuts will be the first to get one with mustard.

24. Use the same sort of reasoning as in Exercise 19, but for sampling *without replacement*, to solve the following problem: If a bank teller has 15 twenty-dollar bills of which three are counterfeits, what is the probability that the eighth of these twenty-dollar bills he gives to a customer is the second of the counterfeit bills he is thus putting into circulation. (*Hint*: first find the probability that one of the first seven bills he hands out is a counterfeit.)

25. The probability that the light of a certain kind of slide projector will last fewer than 40 hours is 0.30, the probability that it will last anywhere from 40 hours to 80 hours is 0.50, and the probability that it will last more than 80 hours is 0.20. What is the probability that among eight such lights two will last fewer than 40 hours, five will last anywhere from 40 to 80 hours, and one will last more than 80 hours?

26. The probability that a certain rifleman hits the bulls-eye is 0.60, the probability that he hits the target but misses the bulls-eye is 0.30, and the probability that he misses the target altogether is 0.10. Assuming independence, what is the probability that in 10 shots he will hit the bulls-eye seven times, the target but not the bulls-eye twice, and miss the target altogether once?

27. Suppose that 60 percent of all state income tax returns filed in a given state are correct, 20 percent contain only mistakes favoring the taxpayer, 10 percent contain only mistakes favoring the state, and 10 percent contain both kinds of mistakes. What is the probability that among 12 of these state income tax returns randomly selected for audit six are correct, three contain only mistakes favoring the taxpayer, one contains only mistakes favoring the state, and two contain both kinds of mistakes?

28. According to the Mendelian theory of heredity, if plants with round yellow seeds are crossbred with plants with wrinkled green seeds, the probabilities of getting a plant that produces round yellow seeds, wrinkled yellow seeds, round green seeds, or wrinkled green seeds are, respectively, $\frac{9}{16}$, $\frac{3}{16}$, $\frac{3}{16}$, and $\frac{1}{16}$. Find the probability that among 11 plants thus obtained there will be five that produce round yellow seeds, two that produce wrinkled yellow seeds, three that produce round green seeds, and one that produces wrinkled green seeds.

29. Using the same sort of reasoning as in the derivation of the formula for the hypergeometric distribution, we can derive a formula which is analogous to the multinomial distribution but applies to *sampling without replacement*. If a set of N objects contains a_1 objects of the first kind, a_2 objects of the second kind, ..., and a_k objects of the kth kind, so that $a_1+a_2+\ldots+a_k = N$, the number of ways in which we can select x_1 objects of the first kind, x_2 objects of the second kind, ..., and x_k objects of the kth kind is given by the *product* of the number of ways in which we can select x_1 of the a_1 objects of the first kind, x_2 of the a_2 objects of the second kind, ..., and x_k of the a_k objects of the kth kind. Then, the probability of obtaining that many objects of each kind is simply this product *divided by* the total number of possibilities, namely, the number of ways in which $x_1+x_2+\ldots+x_k = n$ objects can be selected from the whole set of N objects.

(a) Write a formula for the probability of thus obtaining x_1 objects of the first kind, x_2 objects of the second kind, ..., and x_k objects of the kth kind.

(b) If a library shelf contains 10 detective stories, seven novels, and three non-fiction books, what is the probability that among six books picked at random from this shelf there will be three detective stories, two novels, and one non-fiction book?

30. A panel of prospective jurors consists of seven married men, four single men, six married women, and three single women, and each of them has the same chance of being chosen for a jury of 12. Use the method of Exercise 29 to find the probabilities that the jury will

(a) consist of five married men, one single man, five married women, and one single woman;

(b) consist of four married men, two single men, and six married women?

8

THE LAW OF LARGE NUMBERS

INTRODUCTION

Perhaps, we should have titled this book *How to Live with Uncertainties,* for this is undoubtedly the most important lesson to be learned from the study of probability. There are uncertainties wherever we turn—we cannot be certain that we will like a new job, we cannot be certain that the food we order will taste good, we cannot be certain that a newly designed airplane (or even a regularly scheduled flight) will get off the ground, we cannot be certain that a friendship will last, we cannot be certain that an investment will pay off, and so on, and so on. Undoubtedly, life without uncertainties would be incredibly dull and most uncertainties cannot actually be eliminated, but as we shall see in this concluding chapter, *the vagaries of chance* (*namely, the variations, changes, fluctuations, or differences "caused" by chance*) *are very often measurable, predictable, and to some extent even controllable.*

THE STANDARD DEVIATION

There are many situations in which the difference between *what we expect to get and what we get* can be attributed to chance. This would be the case, for instance, if we got 46 heads in 100 flips of a coin instead of the 50 which we might expect, if a quarterback completes only 12 of 28 forward passes even though his completion rate is normally 50 percent, or if 12 of 500 electric generators coming off an assembly line have to be reassembled even though this is supposed to happen to only 2 percent of them. Of course, this raises the question whether we would still attribute these differences to chance if we got only 16 heads in the 100 flips of the coin, if the quarterback completed only three of 28 passes, or if 63 of the 500 generators had to be reassembled. The answer would probably be "No" in each case, but *where do we draw the line*? Clearly, this would have to depend on *what differences we can reasonably expect* between what we expect and what we get, and since the word "expect" is used here twice, once with regard to the difference and once with regard to the quantity which we observe or predict, this requires some amplification.

If a random variable takes on the value x and its *expected value* (namely, the mean of its distribution) is μ, then the difference $x-\mu$ is called the **deviation from the mean**, and it measures by how much the observed value is off. For example, if we get 46 heads in 100 flips of a balanced coin, then $\mu = 100 \cdot \frac{1}{2} = 50$ and the deviation from the mean is

$$x-\mu = 46-50 = -4$$

Similarly, in the example where 12 of the 500 generators had to be reassembled even though we expected this to happen to only 2 percent, then $\mu = 500(0.02) = 10$ and the deviation from the mean is

$$x-\mu = 12-10 = 2$$

More generally, if a random variable takes on the values $x_1, x_2, \ldots,$ and x_k with the respective probabilities $f(x_1), f(x_2), \ldots,$ and $f(x_k)$, and the mean of its distribution is μ, the deviations from the mean are $x_1-\mu$, $x_2-\mu, \ldots, x_k-\mu$, and *their* (the deviations') expected value is

$$(x_1-\mu)\cdot f(x_1)+(x_2-\mu)\cdot f(x_2)+\ldots+(x_k-\mu)\cdot f(x_k)$$

in accordance with the formula for a mathematical expectation. Hopefully, this quantity should tell us something about the expected size of the chance fluctuations (of the values of the random variable), but unfortunately this is not the case. As the reader will be asked to verify in Exercise 5 on page 209, *the value of the above quantity is always zero, and hence not indicative of anything.* To see what happens, let us refer again to the distribution of the number of heads in three flips of a balanced coin, for which we showed on page 169 that the probabilities for zero, one, two, or three heads are, respectively, $\frac{1}{8}$, $\frac{3}{8}$, $\frac{3}{8}$, and $\frac{1}{8}$, and on page 172 that $\mu = 1\frac{1}{2}$. Thus, the mathematical expectation of the deviations from the mean is

$$\begin{aligned}(0-1\tfrac{1}{2})\cdot\tfrac{1}{8}+(1-1\tfrac{1}{2})\cdot\tfrac{3}{8}+(2-1\tfrac{1}{2})\cdot\tfrac{3}{8}+(3-1\tfrac{1}{2})\cdot\tfrac{1}{8}\\ =(-\tfrac{3}{2})\cdot\tfrac{1}{8}+(-\tfrac{1}{2})\cdot\tfrac{3}{8}+\tfrac{1}{2}\cdot\tfrac{3}{8}+\tfrac{3}{2}\cdot\tfrac{1}{8}\\ =-\tfrac{3}{16}-\tfrac{3}{16}+\tfrac{3}{16}+\tfrac{3}{16}\\ =0\end{aligned}$$

and what happens is that some of the deviations from the mean are positive, some are negative, and their expected value (or average) is *always zero.*

There are several ways in which we can get around this difficulty. Most logical, perhaps, would be to consider only the *size* of the deviations from the mean (namely, ignore their signs or use their *absolute values*), and for the above example we would thus get

$$\frac{3}{2}\cdot\frac{1}{8}+\frac{1}{2}\cdot\frac{3}{8}+\frac{1}{2}\cdot\frac{3}{8}+\frac{3}{2}\cdot\frac{1}{8}=\frac{3}{4}$$

In other words, for three flips of a balanced coin, the number of heads will *on the average* differ (one way or the other) by $\frac{3}{4}$ from the expected value of $\mu = 1\frac{1}{2}$.

Although the method suggested in the preceding paragraph has intuitive appeal, the use of absolute values (namely, ignoring minus signs) leads to all sorts of mathematical complications. It is mainly for this reason that we nearly always work instead with the **squared deviations from the mean**, for the square of a real number is always positive or zero. To illustrate, let us find the expected value of the squared deviations from the mean for the number of heads in three flips of a balanced coin. Using the same value of μ and the same probabilities as above, we get

$$\begin{aligned}(0-1\tfrac{1}{2})^2\cdot\tfrac{1}{8}&+(1-1\tfrac{1}{2})^2\cdot\tfrac{3}{8}+(2-1\tfrac{1}{2})^2\cdot\tfrac{3}{8}+(3-1\tfrac{1}{2})^2\cdot\tfrac{1}{8}\\ &=(-\tfrac{3}{2})^2\cdot\tfrac{1}{8}+(-\tfrac{1}{2})^2\cdot\tfrac{3}{8}+(\tfrac{1}{2})^2\cdot\tfrac{3}{8}+(\tfrac{3}{2})^2\cdot\tfrac{1}{8}\\ &=\tfrac{9}{4}\cdot\tfrac{1}{8}+\tfrac{1}{4}\cdot\tfrac{3}{8}+\tfrac{1}{4}\cdot\tfrac{3}{8}+\tfrac{9}{4}\cdot\tfrac{1}{8}\\ &=\tfrac{3}{4}\end{aligned}$$

and the square root of this quantity, taken to compensate for the fact that we worked with the squared deviations, is $\sqrt{0.75} = 0.87$ (according to Table IV, whose use is explained on page 232.) The fact that the expected value of the absolute values of the deviation from the mean and the expected value of the squared deviations are the same in this example is *purely coincidental*—as the reader will be asked to show in Exercise 4 on page 209, it is *not* the case in general.

To write a general formula for the expected value of the squared deviations from the mean, let us consider again a random variable which takes on the values x_1, x_2, ..., and x_k with the respective probabilities $f(x_1)$, $f(x_2)$, ..., and $f(x_k)$, and whose distribution has the mean μ. Thus, we get

$$(x_1-\mu)^2\cdot f(x_1)+(x_2-\mu)^2\cdot f(x_2)+\ldots+(x_k-\mu)^2\cdot f(x_k)$$

and if we use the $\sum$ notation introduced on page 155, we can also write this quantity, called the **variance** of the distribution, as

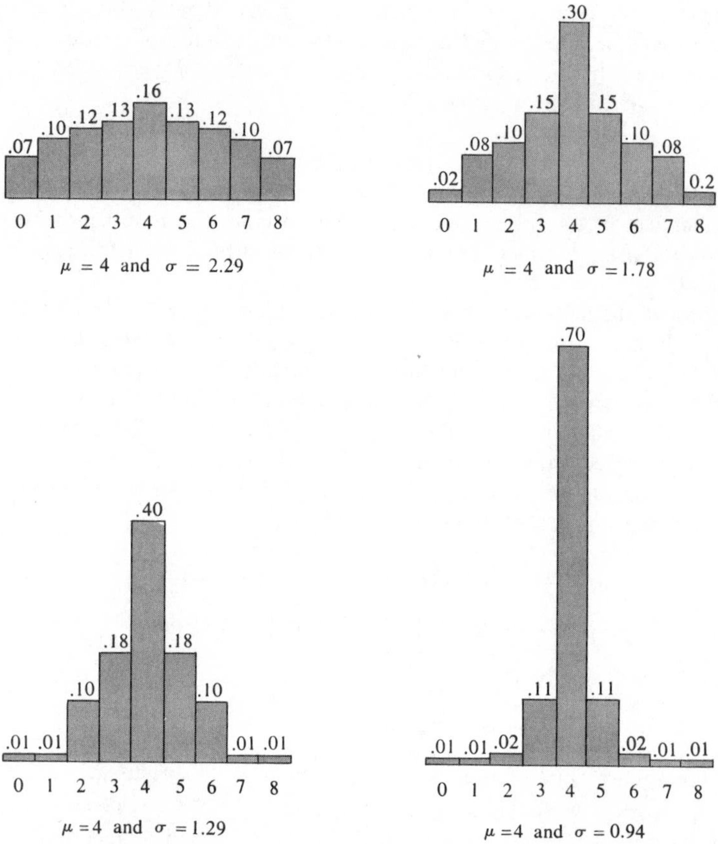

FIGURE 8.1 Distributions with different "spread."

$$\sigma^2 = \sum_{i=1}^{k} (x_i - \mu)^2 \cdot f(x_i)$$

where σ is the lower case Greek letter *sigma*. As we have indicated earlier, we can compensate for working with the squared deviations by taking the square root of the final result, and if we do this now we get the so-called **standard deviation**

$$\sigma = \sqrt{\sum_{i=1}^{k} (x_i - \mu)^2 \cdot f(x_i)}$$

which is probably the most widely used measure of chance variation.

To give the reader some idea of how the standard deviation reflects the average size of chance fluctuations, let us refer to Figure 8.1, which contains the histograms of four probability functions defined for $x = 0, 1, 2, \ldots,$ and 8. They all have the same mean $\mu = 4$ as can easily be verified, but their respective standard deviations are $\sigma = 2.29$, $\sigma = 1.78$, $\sigma = 1.29$, and $\sigma = 0.94$. Aside from the apparent differences in the "spread" of these distributions, the reader can check for himself that the probabilities of getting a value which differs from $\mu = 4$ by *two or more* are, respectively, 0.58, 0.40, 0.24, and 0.08, and that the probabilities of getting a value which differs from $\mu = 4$ by *three or more* are, respectively, 0.34, 0.20, 0.04, and 0.04. [For instance, for the *first* distribution we get

$$(0.07+0.10+0.12)+(0.12+0.10+0.07) = 0.58$$

for values differing from $\mu = 4$ by *two or more,* and a probability of

$$(0.07+0.10)+(0.10+0.07) = 0.34$$

for values differing from $\mu = 4$ by *three or more.*] This serves to illustrate the fact that *when σ is small we are more likely to get a value close to the mean than when σ is large.* This relationship between σ and the probability of getting a value close to the mean is very important, and it will be discussed in detail in the section which follows.

To give another illustration of the calculation of the standard deviation of a probability distribution, let us refer back to the example on page 178, which dealt with the number of letters (among five picked from a large pile) that did not have a ZIP code. Since we already showed on page 180 that the mean of this distribution is approximately 0.5, the work which remains to be done can be arranged as in the following table:

Number of letters x	*Deviation from mean* $x-\mu$	*Squared deviation from mean* $(x-\mu)^2$	*Probability* $f(x)$	$(x-\mu)^2 \cdot f(x)$
0	−0.5	0.25	0.5905	0.1476
1	0.5	0.25	0.3280	0.0820
2	1.5	2.25	0.0729	0.1640
3	2.5	6.25	0.0081	0.0506
4	3.5	12.25	0.0004	0.0049
5	4.5	20.25	0.0000	0.0000
				$\sigma^2 = 0.4491$

This is approximately 0.45, and if we look up the square root of this quantity in Table IV, we get $\sqrt{0.45} = 0.67$.

As in the case of the mean, the calculation of the standard deviation can be greatly simplified when we deal with special distributions. For the *binomial distribution*, for instance, there is the formula

$$\sigma = \sqrt{n \cdot p \cdot (1-p)}$$

which we shall not prove, but which can easily be verified for our two examples. Substituting $n = 3$ and $p = \frac{1}{2}$ into the formula for the number of heads in three flips of a balanced coin, we get

$$\sigma = \sqrt{3 \cdot \tfrac{1}{2} \cdot \tfrac{1}{2}} = \sqrt{0.75} = 0.87$$

and this agrees with the result on page 205. Similarly, substituting $n = 5$ and $p = \frac{1}{10}$ into the formula for the number of letters in five without a ZIP code, we get

$$\sigma = \sqrt{5 \cdot \tfrac{1}{10} \cdot \tfrac{9}{10}} = \sqrt{0.45} = 0.67$$

and this also agrees with the result which we obtained before. As the reader will discover in Exercises 17 and 19 below, there also exist special formulas for the standard deviation of the hypergeometric and geometric distributions.

EXERCISES

1. Referring to Exercise 2 on page 172, find the standard deviation of the distribution of the number of pies which the baker sells on Friday.

2. The following is the probability distribution of the weekly number of bank robberies reported in a Western city:

Number of robberies x	0	1	2	3	4	5	6
Probability $f(x)$	0.272	0.354	0.230	0.100	0.032	0.009	0.003

(a) Find the mean of this distribution.

(b) Use the result of part (a), rounded to one decimal, to find the standard deviation of this distribution.

3. The following is the probability distribution for the number of trout, x, a person will catch in an hour while fishing in the early morning at a certain lake:

x	0	1	2	3	4	5	6	7	8	9
$f(x)$	0.050	0.149	0.224	0.224	0.168	0.101	0.050	0.023	0.008	0.003

(a) Find the mean of this distribution.
(b) Use the result of part (a), rounded to one decimal, to find the standard deviation of this distribution.

4. As can easily be verified, the probabilities of getting zero, one, two, three, four, five, or six heads in six flips of a balanced coin are, respectively, $\frac{1}{64}$, $\frac{6}{64}$, $\frac{15}{64}$, $\frac{20}{64}$, $\frac{15}{64}$, $\frac{6}{64}$, and $\frac{1}{64}$.
(a) Find the mean of this distribution.
(b) Find the expected value of the absolute values of the deviations from the mean.
(c) Find the expected value of the squared deviations from the mean.
(d) Find the standard deviation.
(e) Verify the results of parts (a) and (d) by means of the special formulas for the mean and the standard deviation of a binomial distribution.

Note that the results of parts (b) and (c) are *not* the same.

5. Multiplying out and collecting the respective terms which do and do not have the factor μ, show that the expression on page 204 for the expected value of the deviations from the mean is always zero.

6. Find the standard deviation of the probability distribution on page 169, which pertains to the number of students (among two interviewed) who dislike the new record.

7. If a random variable takes on the values x_1, x_2, ..., and x_k with the respective probabilities $f(x_1)$, $f(x_2)$, ..., and $f(x_k)$, the expected value of the *squares* of its values is given by

$$x_1^2 \cdot f(x_1) + x_2^2 \cdot f(x_2) + \ldots + x_k^2 \cdot f(x_k)$$

and it is usually denoted μ_2 (*mu sub-two*). We have introduced this quantity here because it can be shown that

$$\sigma^2 = \mu_2 - \mu^2$$

namely, that the variance is μ_2 *minus* the square of the mean. This is important because we can thus calculate σ^2, and hence σ, *without having to bother with the deviations from the mean.* For instance, for the example on page 169 (which dealt with the number of heads in three flips of a coin), we can now write directly

$$\mu_2 = 0^2 \cdot \tfrac{1}{8} + 1^2 \cdot \tfrac{3}{8} + 2^2 \cdot \tfrac{3}{8} + 3^2 \cdot \tfrac{1}{8} = \tfrac{24}{8} = 3$$

so that $\sigma^2 = 3 - (\frac{3}{2})^2 = 3 - \frac{9}{4} = \frac{3}{4}$. Use this method to verify the value obtained for σ in

(a) Exercise 2; (c) Exercise 4;
(b) Exercise 3; (d) Exercise 6.

8. Verify the formula $\sigma^2 = \mu_2 - \mu^2$ of the preceding exercise by multiplying out the squares in the expression for the variance on page 206 and collecting terms. [*Hint*: $(x_1-\mu)^2$, for example, equals $x_1^2 - 2x_1\mu + \mu^2$.]

9. Referring to the table on page 176, which pertains to the number of cars requiring repairs within the first 90 days, find the standard deviation by using
(a) either the original formula on page 206 or the short-cut formula for σ^2 in Exercise 7;
(b) the special formula for the standard deviation of a binomial distribution.

10. Find the standard deviation of the distribution of Exercise 20 on page 187 using
(a) the original formula for σ on page 206;
(b) the short-cut formula for σ^2 in Exercise 7;
(c) the special formula for the standard deviation of a binomial distribution.

11. Find the standard deviation of the binomial distribution of Exercise 21 on page 187 using
(a) either the original formula on page 206 or the short-cut formula for σ^2 in Exercise 7;
(b) the special formula on page 208.

12. Use the special formula for the standard deviation of a binomial distribution to find σ for each part of Exercise 22 on page 188.

13. With reference to the example on page 140 where a person randomly grabs two of 15 mystery books, four of which he has already read, we have already shown that the probabilities of his getting zero, one, or two books which he has already read are, respectively, $\frac{11}{21}$, $\frac{44}{105}$, and $\frac{2}{35}$, and that the mean of this distribution is $\frac{8}{15}$. Find the standard deviation of this distribution using
(a) the original formula for σ on page 206;
(b) the short-cut formula for σ^2 in Exercise 7.

14. Referring to the results of Exercises 2 and 3 on page 197 and using the original formula for σ on page 206, find the standard deviation of the distribution of the number of defective toasters.

15. Referring to the results of Exercises 5 and 6 on page 198 and using the short-cut formula for σ^2 in Exercise 7, find the standard deviation of the distribution of the number of air mail letters with the right kind of stamp.

16. Referring to the results of Exercises 7 and 8 on page 198 and using the original formula for σ on page 206, find the standard deviation of the distribution of the number of counterfeits.

17. Using the same notation as on page 191, we can write a special formula for the standard deviation of the hypergeometric distribution as

$$\sigma = \sqrt{\frac{nab(a+b-n)}{(a+b)^2(a+b-1)}}$$

Use this formula to verify the values obtained for σ in

(a) Exercise 13;
(b) Exercise 14;
(c) Exercise 15;
(d) Exercise 16.

18. Use the results of Exercise 18 on page 199 to find the standard deviation of the distribution of the number of the trial on which a balanced coin comes up heads for the first time.

19. Compare the result of Exercise 18 with the *exact* result which is given by

$$\sigma = \sqrt{\frac{1-p}{p^2}}$$

where p is the constant probability of success for each trial.

CHEBYSHEV'S THEOREM

In the preceding section we introduced the standard deviation as a measure of the *expected chance fluctuations* of a random variable, and as we already pointed out on page 207, *when σ is small we are more likely to get a value which is close to the mean than when σ is large.* Formally, this important idea is expressed by the following rule, called **Chebyshev's Theorem** (after the nineteenth-century Russian mathematician P. L. Chebyshev):

> **The probability that a random variable will take on a value within k standard deviations of the mean is at least $1-\frac{1}{k^2}$, where k can be any positive number.**

Thus, the probability of getting a value within *two* standard deviations of the mean (namely, between $\mu-2\sigma$ and $\mu+2\sigma$) is always *at least* $1-\frac{1}{2^2}=0.75$, and the probability of getting a value within *three* standard deviations of the mean (namely, on the interval from $\mu-3\sigma$ to $\mu+3\sigma$) is always *at least* $1-\frac{1}{3^2}=0.89$. Also, the probability of getting a value within *five* standard deviations of the mean is always *at least* $1-\frac{1}{5^2}=0.96$, and the probability of getting a value within *ten* standard deviations of the mean is always *at least* $1-\frac{1}{10^2}=0.99$.

If we change the argument around, Chebyshev's theorem can also be stated in the following form:

> **The probability that a random variable will take on a value which differs from the mean by at least k standard deviations is at most $\frac{1}{k^2}$.**

Thus, the probability of getting a value which differs from the mean by at least *two* standard deviations is *at most* $\frac{1}{2^2}=0.25$, the probability of getting a value which differs from the mean by at least *four* standard deviations is *at most* $\frac{1}{4^2}=0.06$, and the probability of getting a value which differs from the mean by at least *ten* standard deviations is *at most* $\frac{1}{10^2}=0.01$. Of course, when we say "differs" here, we mean "differs either way." The proof of Chebyshev's theorem is not too difficult, but it is not essential for an understanding of what is involved; thus, it is left to the end of the chapter, where it is given in a Technical Note.

To consider some concrete applications, suppose that the number of telephone calls which a doctor receives between 9 A.M. and 10 A.M. is a random variable, whose distribution has the mean $\mu = 18$ and the standard deviation $\sigma = 4.24$. For example, using Chebyshev's theorem with $k = 2$ we can say that the probability is *at least* 0.75 (or that the odds are *at least* 3 to 1) that he will receive anywhere between $18-2(4.24) = 9.52$ and $18+2(4.24) = 26.48$ calls, namely, that he will receive from 10 to 26 calls. Also, for $k = 10$ we can say that the probability is *at most* 0.01 that the number of calls will differ from 18 by *at least* $10(4.24) = 42.40$, namely, that he will receive 61 or more calls. (We did not have to mention the lower limit in this case as it is a negative number.)

To give an example which pertains to the binomial distribution, suppose that an insurance company which writes industrial accident insurance figures that 30 percent of its policy holders will file at least one claim per year. *Should they suspect that something may be wrong if in a given year 2,292 of its 6,200 policy holders in Massachussetts file at least one claim?* Since the mean and the standard deviation of the corresponding distribution are

$$\mu = 6{,}200(0.30) = 1{,}860$$

and

$$\sigma = \sqrt{6{,}200(0.30)(0.70)} = \sqrt{1{,}302} = 36$$

we find that the amount by which that year's figure differs from the mean, namely, $2{,}292 - 1{,}860 = 432$, is equal to $\frac{432}{36} = 12$ standard deviations. Since the probability of being off by that much or more is *at most* $\frac{1}{12^2} = 0.007$ according to Chebyshev's theorem, it stands to reason that the insurance company might question the "30 percent figure" on which they based their expectations.

THE LAW OF LARGE NUMBERS

When Chebyshev's theorem is applied to random variables having binomial distributions, it leads to the so-called **Law of Large Numbers.** For instance, for the distribution of the number of heads in 400 flips of a balanced coin we have $\mu = 400 \cdot \frac{1}{2} = 200$ and $\sigma = \sqrt{400 \cdot \frac{1}{2} \cdot \frac{1}{2}} = 10$, and we can thus make the following assertion according to Chebyshev's theorem with $k = 5$: *The probability is at least* $1 - \frac{1}{5^2} = 0.96$ *that in 400 flips of a balanced coin the number of heads will differ from* $\mu = 200$ *by less than* $5(10) = 50$. Expressing this result in terms of proportions (namely, dividing 200 as well as 50 by $n = 400$), we find that *the probability is at least 0.96 that in 400 flips of a balanced coin the proportion of heads will differ from* $\frac{200}{400} = \frac{1}{2}$ *by less than* $\frac{50}{400} = \frac{1}{8}$ *or* 0.125. To continue with this example, the reader will be asked to show in Exercise 8 on page 216 that the probability is at least 0.96 that in 10,000 flips of a balanced coin the proportion of heads will differ from $\frac{1}{2}$ by less than 0.025, and that the probability is at least 0.9996 that in 1,000,000 flips of a balanced coin the proportion of heads will differ from $\frac{1}{2}$ by less than 0.025. Note that by increasing the number of flips we can thus "shrink" the interval for the proportion and/or increase the probability with which we make the assertion.

In general, we can thus assert for a random variable having the binomial distribution that *the probability is at least* $1-\frac{1}{k^2}$ *that the number of successes in n trials will differ from* $\mu = n \cdot p$ *by less than* $k\sigma = k \cdot \sqrt{np(1-p)}$. Then, if we divide by n, as above, to express the result in terms of proportions, we arrive at the following, called the **Law of Large Numbers:**

> **If a random variable has the binomial distribution, the probability is at least** $1-\frac{1}{k^2}$ **that the proportion of successes in** n **trials will differ from** p **by less than** $k \cdot \sqrt{\frac{p(1-p)}{n}}$.

Note that in the expression for the difference, $k \cdot \sqrt{\frac{p(1-p)}{n}}$, n appears only in the denominator, which means that *we can make this quantity as small as we want by making n sufficiently large.* We can do this regardless of whether $k = 5$, $k = 10$, $k = 100$, or perhaps even $k = 1{,}000$, and $1-\frac{1}{k^2}$ is correspondingly close to 1. Thus, the Law of Large Numbers tells us that

> **When** n **is sufficiently large, we can be virtually certain that the proportion of successes will be very close (as close as we want) to the probability** p **of success for each trial.**

To give a numerical example, suppose we are concerned about the possibility that after rolling a pair of dice 14,000 times the observed proportion of *12's* may still differ from the probability of rolling a 12, namely, $\frac{1}{36}$, by 0.01 or more. To find the probability that the proportion of 12's will differ from $\frac{1}{36}$ by less than 0.01, we first put 0.01 equal to $k \cdot \sqrt{\frac{p(1-p)}{n}}$ with $n = 14{,}000$ and $p = \frac{1}{36}$, and solve for k. We thus get

$$0.01 = k \cdot \sqrt{\frac{\frac{1}{36} \cdot \frac{35}{36}}{14{,}000}}$$

which gives $k = 7.2$, and we can assert that *the probability is at least* $1-\frac{1}{(7.2)^2} = 0.98$ *that after 14,000 rolls of a pair of balanced dice the observed proportion of twelves will differ from* $\frac{1}{36}$ *by less than 0.01.* Correspondingly, *the probability is at most* $1-0.98 = 0.02$ *that the observed proportion of* *twelves will differ from $\frac{1}{36}$ by 0.01 or more.* If this is not enough, we simply have to roll the dice a few more times, and in Exercise 11 on page 216 the reader will be asked to show that for 56,000 rolls of the dice the probability is at least 0.995 (instead of 0.98) that the observed proportion of 12's will differ from $\frac{1}{36}$ by less than 0.01.

Actually, the Law of Large Numbers is of great theoretical significance, but we have given it here mainly because it provides the basis for the *frequency interpretation* of probability which we studied in Chapter 2—after all, it was there that we talked about what happens to a proportion "in the long run." Of course, the Law of Large Numbers is a rule which follows from the Postulates of Probability and it applies regardless of what "probability yardstick" of Chapter 2 we happen to use. In fact, it tells us that probabilities—whatever they may be—can be estimated in terms of observed proportions. The only difficulty is that this argument applies only to situations which can, in some sense, be repeated, and this takes us back to the discussion of Chapter 2.

EXERCISES

1. A student answers the 64 questions of a true-false test by flipping a balanced coin (*heads* is "true" and *tails* is "false").
(a) Use the special formulas on pages 180 and 208 to calculate μ and σ for the distribution of the number of correct answers he will get.
(b) What does Chebyshev's theorem with $k = 3$ tell us about the number of correct answers he should get?
(c) According to Chebyshev's theorem, with what probability can we assert that he will receive more than 16 but fewer than 48 correct answers?

2. The annual number of rainy days in a city is a random variable with $\mu = 144$ and $\sigma = 12$. According to Chebyshev's theorem, with what probability can we assert that in any given year there will be
(a) more than 84 but fewer than 204 rainy days;
(b) at least 174 rainy days or at most 114 rainy days?

3. The number of marriage licenses issued in a certain city during the month of June averages $\mu = 124$ with a standard deviation of $\sigma = 8.5$.
(a) According to Chebyshev's theorem with $k = 4$, what can we assert about the number of marriage licenses that should be issued in this city during the month of June?
(b) According to Chebyshev's theorem, with what probability can we assert that there will be at least 175 marriage licenses issued in this city during the month of June?

4. According to Chebyshev's theorem, the probability of getting a value which differs from the mean by at least two standard deviations is at most 0.25. What is the corresponding *exact* probability for a binomial distribution with $n = 9$ and $p = \frac{1}{2}$? (*Hint*: find the values of μ and σ with the special formulas, calculate $\mu - 2\sigma$ and $\mu + 2\sigma$, and look up the necessary probabilities in Table III.)

5. Referring to Exercise 2 on page 208 and using Chebyshev's theorem with $k = 3.3$, what can we assert about the number of bank robberies in that city in any given week?

6. Referring to Exercise 3 on page 208 and using Chebyshev's theorem with $k = 6$, what can we assert about the number of trout a person should be able to catch while fishing for an hour at the given lake?

7. What does Chebyshev's theorem tell us about the probability of getting more than 30 but fewer than 105 *fours* in 405 rolls of a balanced die?

8. Verify the statements made on page 213 about the number of heads in 10,000 and 1,000,000 flips of a balanced coin.

9. Use Chebyshev's theorem to verify that the probability is at least $\frac{35}{36}$ that
(a) in 900 flips of a balanced coin the proportion of heads will differ from 0.50 by less than 0.10;
(b) in 3,600 flips of a balanced coin the proportion of heads will differ from 0.50 by less than 0.05;
(c) in 90,000 flips of a balanced coin the proportion of heads will differ from 0.50 by less than 0.01.

10. Use Chebyshev's theorem to verify that the probability is at least $\frac{15}{16}$ that for a random variable having a binomial distribution with $p = \frac{1}{4}$
(a) the proportion of successes in 300 trials will differ from 0.25 by less than 0.10;
(b) the proportion of successes in 7,500 trials will differ from 0.25 by less than 0.02;
(c) the proportion of successes in 120,000 trials will differ from 0.25 by less than 0.005.

11. Verify the assertion on page 215 that when $n = 56{,}000$, the probability is at least 0.995 that the proportion of *twelves* rolled with a pair of dice will differ from $\frac{1}{36}$ by less than 0.01.

12. With what probability can we assert that in 10,000 flips of a balanced coin the proportion of heads will differ from 0.50 by less than 0.025?

13. The number of trials needed to be able to assert with a *given probability* that the difference between the proportion of successes and the probability p (of success for each trial) is less than a *given quantity* c, can be obtained by solving for n in the equation

$$c = k\cdot\sqrt{\frac{p(1-p)}{n}}$$

Here c and p are given directly and k is obtained by equating the given probability to $1-\frac{1}{k^2}$ and solving for k. For instance, if we want to be able to assert with a probability of at least $\frac{15}{16}$ that the proportion of 5's we get with a balanced die will differ from $p=\frac{1}{6}$ by less than $c=\frac{1}{60}$, we must first solve the equation $1-\frac{1}{k^2}=\frac{15}{16}$. This gives $k=4$, and if we substitute this value together with $p=\frac{1}{6}$ and $c=\frac{1}{60}$ into the above equation, we get

$$\frac{1}{60} = 4\cdot\sqrt{\frac{\frac{1}{6}\cdot\frac{5}{6}}{n}}$$

Show that the solution of this equation is $n = 8{,}000$ and, hence that the die has to be rolled 8,000 times to enable us to assert with a probability of at least $\frac{15}{16}$ that the proportion of 5's will differ from $\frac{1}{6}$ by less than $\frac{1}{60}$.

14. Use the method of the preceding exercise to determine how many times we have to flip a balanced coin to be able to assert with a probability of at least 0.99 that the difference between the proportion of *tails* and 0.50 is less than 0.04.

TECHNICAL NOTE

To prove Chebyshev's theorem, let us consider a random variable which takes on the values $x_1, x_2, \ldots$, and x_k with the respective probabilities $f(x_1), f(x_2), \ldots$, and $f(x_k)$, and whose distribution has the mean μ and the variance

$$\sigma^2 = \sum_{i=1}^{k} (x_i-\mu)^2\cdot f(x_i)$$

Then, let us subdivide (or partition) this sum into *three parts* as indicated in Figure 8.2, and write*

$$\sigma^2 = \sum_1 (x_i-\mu)^2\cdot f(x_i) + \sum_2 (x_i-\mu)^2\cdot f(x_i) + \sum_3 (x_i-\mu)^2\cdot f(x_i)$$

* This kind of diagram is called a **bar chart**, and it differs from a histogram like that of Figure 7.2 in that the rectangles (or bars) are narrower; the probabilities are still represented by the heights.

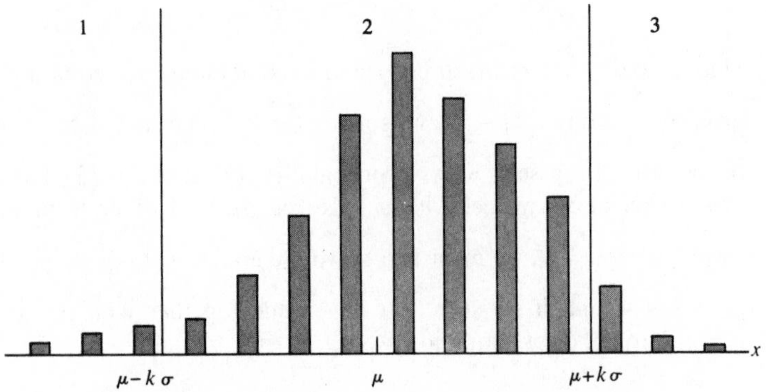

FIGURE 8.2 Bar chart for proof of Chebyshev's theorem.

where $\sum_1$ is summed over all values of the random variable which are less than or equal to $\mu-k\sigma$, $\sum_2$ is summed over all values of the random variable which are between $\mu-k\sigma$ and $\mu+k\sigma$, and $\sum_3$ is summed over all values of the random variable which are greater than or equal to $\mu+k\sigma$. Note that in $\sum_1$ the deviations from the mean $x_i-\mu$ are less than or equal to $-k\sigma$, in $\sum_2$ they are greater than $-k\sigma$ but less than $k\sigma$, and in $\sum_3$ they are greater than or equal to $k\sigma$.

Since the quantities $(x_i-\mu)^2 \cdot f(x_i)$ are all *positive or zero*—the first factor is a square and the second a probability—it follows that $\sum_2 (x_i-\mu)^2 \cdot f(x_i)$ is positive or zero, and if we subtract this quantity from the above expression for σ^2, the expression is either reduced or remains the same. Symbolically, we can express this by writing

$$\sigma^2 \geqq \sum_1 (x_i-\mu)^2 \cdot f(x_i) + \sum_3 (x_i-\mu)^2 \cdot f(x_i)$$

where the inequality sign serves to indicate that the expression on the right is *less than or equal to* σ^2. Now, observe that in the first and third sums we included only those values of the random variable which differed from the mean by at least $k\sigma$, namely, those values for which $x_i-\mu$ is at least $k\sigma$ in *magnitude*, or absolute value, and hence $(x_i-\mu)^2$ is at least $(k\sigma)^2 = k^2\sigma^2$. Thus, if we substitute $k^2\sigma^2$ for $(x_i-\mu)^2$ in both sums of the above inequality, we may be making the expression on the righthand side smaller yet, and we can write

$$\sigma^2 \geqq \sum_1 k^2\sigma^2 \cdot f(x_i) + \sum_3 k^2\sigma^2 \cdot f(x_i)$$

Finally, since k^2 and σ^2 are always positive (except for the very trivial case where a random variable takes on only one value and $\sigma = 0$, which we shall exclude), we can divide the expressions on both sides of the last inequality by $k^2\sigma^2$, and get

$$\frac{1}{k^2} \geqq \sum_1 f(x_i) + \sum_3 f(x_i)$$

This actually completes the proof of Chebyshev's theorem in the second form in which it was given on page 212. Since $\sum_1 f(x_i)$ is the sum of the probabilities associated with all values of the random variable that are less than or equal to $\mu - k\sigma$ and $\sum_3 f(x_1)$ is the sum of the probabilities associated with all values of the random variable that are greater than or equal to $\mu + k\sigma$, it follows that $\sum_1 f(x_i) + \sum_3 f(x_i)$ is the probability that the random variable will take on a value which differs from μ by at least $k\sigma$, and we have thus shown that this probability is at most $\frac{1}{k^2}$.

SUMMARY

The title of a textbook seldom gives a complete picture of its contents, and the one which we have chosen for this book is no exception. Not only can the subject of probability be introduced at various levels of mathematical refinement, but it can also be presented with an emphasis on questions of meaning, with an emphasis on the mathematical theory of probability, with an emphasis on applications, or with an emphasis on any two or all three of these aspects. The subject of probability is presented in this book at the lowest level of mathematical rigor at which, in the opinion of the author, it can effectively be taught. Also, it is designed to present a challenging and stimulating balance between questions concerning the meaning of probability statements, questions concerning the determination (estimation) of probabilities, and questions concerning their mathematical "manipulation." As it is presented here, the study of probability also presents an immediate application of the set concepts, which form such an important part of modern mathematics.

So far as applications are concerned, the illustrations and exercises in this book were chosen mostly from situations close to everyday life, with the hope that this will impress upon the reader the *relevance* of the study of probability. A possible disadvantage of this approach may be that we have not sufficiently impressed upon the reader the role played by probabilities in scientific thought. As we pointed out in the Introduction, it is a misconception to think that probabilities are needed only as a cover-up for ignorance, namely, that probabilities would not be needed if only we took the time to acquire enough information. Indeed, *in modern science, probabilities constitute an integral part of many laws of nature*. This is illustrated by the following examples: In the *Mendelian theory of heredity*, to which we referred in Exercise 28 on page 201, the inheritance of characteristics such as the shape or color of plants (or, say, color-blindness in humans) follows laws which can be described only in terms of probabilities. Also, in modern theoretical physics the *Heisenberg indeterminance principle* tells us that the more certain we are about the position (or location) of a particle, the less certain we will be about its speed (or momentum). Note that this actually *is* a law of nature, and not merely a reflection of the shortcomings of the instruments with which we measure. Even the flow of heat is controlled by laws which, strictly speaking, can be expressed only in probabilistic terms. Here we can be "practically certain" that *the most likely thing will happen*, and if we hold one end of a metal rod over a fire, the most likely thing is that the heat will distribute itself evenly over the whole rod; the least likely is for one end of the rod to remain very hot and the other to remain cold.

Returning to the question of relevance, a very important reason for the study of probability is its role in *statistics*, which has virtually become a required subject for all courses of study in the biological, physical, and social sciences. The reason for this is the increasingly *quantitative approach* employed in these sciences (as well as in business and many other activities which directly affect our lives). Since most of the information required by this approach comes from *samples*, its analysis involves uncertainties and, hence, can be evaluated only in terms of probabilities. Indeed, modern statistics is often defined as the art, or science, of "decision making in the face of uncertainty," and hence comes under the heading of applied probability theory. What the reader has learned in this book will provide him with a valuable background for the study of statistics, and it should be pointed out that quite a few of our illustrations and exercises, for instance, the example dealing with the inspection of the cameras on page 189 and Exercise 15 on page 186, are actually statistical decision problems.

Needless to say, statistics is not the only area where probability theory finds applications. Very important also is the study of **random processes** where chance events take place (repeatedly, perhaps) in the course of time. This includes repeated flips of a coin or rolls of a pair of dice, but more significantly it includes such diverse phenomena as telephone calls arriving at a switchboard, queues (waiting lines) forming in a bank, radiation recorded by means of a Geiger counter, the spreading of epidemics (or rumors), cars entering a freeway, the vibration of airplane wings, the fluctuating production output of an assembly line, or imperfections in a roll of cloth, to mention but a few. Due to the very nature of these phenomena, the *mathematical models* (methods, or theories) which we use in their study have to center on probabilities.

To conclude this Summary, let us return briefly to the remarks which we made in Chapter 3 about the art of *divination* (namely, the practice of seeking answers from the supernatural through games of chance). In recent years, the practice of basing decisions on the outcomes of chance happenings has become quite respectable through the use of so-called **Monte Carlo methods**; we should add, though, that there is no longer any question of divine intervention or divine intent. Monte Carlo methods are essentially **simulation techniques**, and they enable us to study in the classroom or in the laboratory random processes which would otherwise be difficult to observe. They have been used, for example, to study the effect of changes in an assembly procedure without actually having to put the changes into operation, and the effects of pollution without having to induce them in our environment. Very often, the use of Monte Carlo methods eliminates the cost of building and operating expensive physical equipment; it is thus used in the study of collisions of photons with electrons, the scattering of neutrons, and other complicated phenomena. The chance factors in all these processes are simulated by means of appropriate gambling devices, which, most of the time, are themselves simulated by means of electronic computers. (For instance, a computer could be used to simulate repeated flips of a coin by picking one-digit numbers from a vast array of numbers in its storage, letting 0, 2, 4, 6, and 8 represent *heads,* and 1, 3, 5, 7, and 9 *tails.*)

A classical example of the use of a gambling device in solving a problem of pure mathematics is the following determination of π (the ratio of the circumference of a circle to its diameter). Early in the eighteenth century, the French naturalist George de Buffon showed that if a very fine needle of length a is thrown at random on a board ruled with equidistant parallel lines, the probability that the needle will intersect one of the lines is $\frac{2a}{\pi b}$, where b is the distance between the parallel lines. The remarkable thing about this result is that it involves the constant $\pi = 3.1415926\ldots$, which can thus be *estimated* by actually tossing a needle on a board suitably ruled with parallel lines. Early experiments of this kind performed in the middle of the nineteenth century yielded an estimate of 3.155 (based on 3,204 trials) and an estimate of 3.1519 (based on 5,000 trials). Note that a *probabilistic approach* has been used here to solve a *non-probabilistic* problem, and this has taken us far from the examples on page 78, where other non-probabilistic problems were solved by interpreting chance happenings as divine intent. "She loves me, she loves me not, she loves me, she loves me not, . . ."

BIBLIOGRAPHY

Borel, E., *Elements of the Theory of Probability.* Englewood Cliffs, N.J.: Prentice-Hall, 1965.

Burford, R. L., *Introduction to Finite Probability.* Columbus, Ohio: Charles E. Merrill Books, 1967.

David, F. N., *Games, Gods and Gambling.* New York: Hafner Publishing Co., 1962.

David, F. N., and Barton, D. E., *Combinatorial Chance.* New York: Hafner Publishing Co., 1962.

Dixon, J. R., *A Programmed Introduction to Probability.* New York: John Wiley & Sons, 1964.

Feller, W., *An Introduction to Probability Theory and its Applications, Vol. I.* New York: John Wiley & Sons, 1957.

Freund, J. E., *Mathematical Statistics, 2nd ed.* Englewood Cliffs, N.J.: Prentice-Hall, 1971.

Goldberg, S., *Probability—An Introduction.* Englewood Cliffs, N.J.: Prentice-Hall, 1960.

Good, I. J., *The Estimation of Probabilities: An Essay on Modern Bayesian Methods.* Cambridge, Mass.: Massachusetts Institute of Technology Press, 1965.

Hodges, J. L., and Lehmann, E. L., *Elements of Finite Probability.* San Francisco: Holden-Day, 1965.

Hoel, P., *Introduction to Mathematical Statistics, 4th ed.* New York: John Wiley & Sons, 1971.

Morgan, B. W., *An Introduction to Bayesian Statistical Decision Processes.* Englewood Cliffs, N.J.: Prentice-Hall, 1968.

Mosteller, F., Rourke, R. E. K., and Thomas, G. B., *Probability with Statistical Applications.* Reading Mass.: Addison-Wesley, 1961.

Nagel, E., *Principles of the Theory of Probability.* Chicago: University of Chicago Press, 1939.

National Bureau of Standards, *Tables of the Binomial Probability Distribution, Applied Mathematics Series No. 6.* Washington, D. C.: U. S. Government Printing Office, 1950.

Parzen, E., *Modern Probability Theory and Its Applications*. New York: John Wiley & Sons, 1960.

Reichenbach, H., *The Theory of Probability*. Berkeley, Calif.: University of California Press, 1949.

Romig, H. G., *50–100 Binomial Tables*. New York: John Wiley & Sons, 1953.

Thorp, E. O., *Elementary Probability*. New York: John Wiley & Sons, 1966.

Whitworth, W. A., *Choice and Chance, 5th ed*. New York: Hafner Publishing Co., 1959.

TABLES

I. Factorials

II. Binomial Coefficients

III. Binomial Probabilities

IV. Square Roots

TABLE I

FACTORIALS

n	$n!$
0	1
1	1
2	2
3	6
4	24
5	120
6	720
7	5,040
8	40,320
9	362,880
10	3,628,800
11	39,916,800
12	479,001,600
13	6,227,020,800
14	87,178,291,200
15	1,307,674,368,000

TABLE II

BINOMIAL COEFFICIENTS*

n	$\binom{n}{0}$	$\binom{n}{1}$	$\binom{n}{2}$	$\binom{n}{3}$	$\binom{n}{4}$	$\binom{n}{5}$	$\binom{n}{6}$	$\binom{n}{7}$	$\binom{n}{8}$	$\binom{n}{9}$	$\binom{n}{10}$
0	1										
1	1	1									
2	1	2	1								
3	1	3	3	1							
4	1	4	6	4	1						
5	1	5	10	10	5	1					
6	1	6	15	20	15	6	1				
7	1	7	21	35	35	21	7	1			
8	1	8	28	56	70	56	28	8	1		
9	1	9	36	84	126	126	84	36	9	1	
10	1	10	45	120	210	252	210	120	45	10	1
11	1	11	55	165	330	462	462	330	165	55	11
12	1	12	66	220	495	792	924	792	495	220	66
13	1	13	78	286	715	1287	1716	1716	1287	715	286
14	1	14	91	364	1001	2002	3003	3432	3003	2002	1001
15	1	15	105	455	1365	3003	5005	6435	6435	5005	3003
16	1	16	120	560	1820	4368	8008	11440	12870	11440	8008
17	1	17	136	680	2380	6188	12376	19448	24310	24310	19448
18	1	18	153	816	3060	8568	18564	31824	43758	48620	43758
19	1	19	171	969	3876	11628	27132	50388	75582	92378	92378
20	1	20	190	1140	4845	15504	38760	77520	125970	167960	184756

* If necessary, use the formula $\binom{n}{k} = \binom{n}{n-k}$.

TABLE III

BINOMIAL PROBABILITIES*

		p										
n	x	0.05	0.1	0.2	0.3	0.4	0.5	0.6	0.7	0.8	0.9	0.95
2	0	0.902	0.810	0.640	0.490	0.360	0.250	0.160	0.090	0.040	0.010	0.002
	1	0.095	0.180	0.320	0.420	0.480	0.500	0.480	0.420	0.320	0.180	0.095
	2	0.002	0.010	0.040	0.090	0.160	0.250	0.360	0.490	0.640	0.810	0.902
3	0	0.857	0.729	0.512	0.343	0.216	0.125	0.064	0.027	0.008	0.001	
	1	0.135	0.243	0.384	0.441	0.432	0.375	0.288	0.189	0.096	0.027	0.007
	2	0.007	0.027	0.096	0.189	0.288	0.375	0.432	0.441	0.384	0.243	0.135
	3		0.001	0.008	0.027	0.064	0.125	0.216	0.343	0.512	0.729	0.857
4	0	0.815	0.656	0.410	0.240	0.130	0.062	0.026	0.008	0.002		
	1	0.171	0.292	0.410	0.412	0.346	0.250	0.154	0.076	0.026	0.004	
	2	0.014	0.049	0.154	0.265	0.346	0.375	0.346	0.265	0.154	0.049	0.014
	3		0.004	0.026	0.076	0.154	0.250	0.346	0.412	0.410	0.292	0.171
	4			0.002	0.008	0.026	0.062	0.130	0.240	0.410	0.656	0.815
5	0	0.774	0.590	0.328	0.168	0.078	0.031	0.010	0.002			
	1	0.204	0.328	0.410	0.360	0.259	0.156	0.077	0.028	0.006		
	2	0.021	0.073	0.205	0.309	0.346	0.312	0.230	0.132	0.051	0.008	0.001
	3	0.001	0.008	0.051	0.132	0.230	0.312	0.346	0.309	0.205	0.073	0.021
	4			0.006	0.028	0.077	0.156	0.259	0.360	0.410	0.328	0.204
	5				0.002	0.010	0.031	0.078	0.168	0.328	0.590	0.774
6	0	0.735	0.531	0.262	0.118	0.047	0.016	0.004	0.001			
	1	0.232	0.354	0.393	0.303	0.187	0.094	0.037	0.010	0.002		
	2	0.031	0.098	0.246	0.324	0.311	0.234	0.138	0.060	0.015	0.001	
	3	0.002	0.015	0.082	0.185	0.276	0.312	0.276	0.185	0.082	0.015	0.002
	4		0.001	0.015	0.060	0.138	0.234	0.311	0.324	0.246	0.098	0.031
	5			0.002	0.010	0.037	0.094	0.187	0.303	0.393	0.354	0.232
	6				0.001	0.004	0.016	0.047	0.118	0.262	0.531	0.735
7	0	0.698	0.478	0.210	0.082	0.028	0.008	0.002				
	1	0.257	0.372	0.367	0.247	0.131	0.055	0.017	0.004			
	2	0.041	0.124	0.275	0.318	0.261	0.164	0.077	0.025	0.004		
	3	0.004	0.023	0.115	0.227	0.290	0.273	0.194	0.097	0.029	0.003	
	4		0.003	0.029	0.097	0.194	0.273	0.290	0.227	0.115	0.023	0.004

* All values omitted in this table are 0.0005 or less.

TABLE III

BINOMIAL PROBABILITIES (*Continued*)

		p										
n	*x*	0.05	0.1	0.2	0.3	0.4	0.5	0.6	0.7	0.8	0.9	0.95
7	5			0.004	0.025	0.077	0.164	0.261	0.318	0.275	0.124	0.041
	6				0.004	0.017	0.055	0.131	0.247	0.367	0.372	0.257
	7					0.002	0.008	0.028	0.082	0.210	0.478	0.698
8	0	0.663	0.430	0.168	0.058	0.017	0.004	0.001				
	1	0.279	0.383	0.336	0.198	0.090	0.031	0.008	0.001			
	2	0.051	0.149	0.294	0.296	0.209	0.109	0.041	0.010	0.001		
	3	0.005	0.033	0.147	0.254	0.279	0.219	0.124	0.047	0.009		
	4		0.005	0.046	0.136	0.232	0.273	0.232	0.136	0.046	0.005	
	5			0.009	0.047	0.124	0.219	0.279	0.254	0.147	0.033	0.005
	6			0.001	0.010	0.041	0.109	0.209	0.296	0.294	0.149	0.051
	7				0.001	0.008	0.031	0.090	0.198	0.336	0.383	0.279
	8					0.001	0.004	0.017	0.058	0.168	0.430	0.663
9	0	0.630	0.387	0.134	0.040	0.010	0.002					
	1	0.299	0.387	0.302	0.156	0.060	0.018	0.004				
	2	0.063	0.172	0.302	0.267	0.161	0.070	0.021	0.004			
	3	0.008	0.045	0.176	0.267	0.251	0.164	0.074	0.021	0.003		
	4	0.001	0.007	0.066	0.172	0.251	0.246	0.167	0.074	0.017	0.001	
	5		0.001	0.017	0.074	0.167	0.246	0.251	0.172	0.066	0.007	0.001
	6			0.003	0.021	0.074	0.164	0.251	0.267	0.176	0.045	0.008
	7				0.004	0.021	0.070	0.161	0.267	0.302	0.172	0.063
	8					0.004	0.018	0.060	0.156	0.302	0.387	0.299
	9						0.002	0.010	0.040	0.134	0.387	0.630
10	0	0.599	0.349	0.107	0.028	0.006	0.001					
	1	0.315	0.387	0.268	0.121	0.040	0.010	0.002				
	2	0.075	0.194	0.302	0.233	0.121	0.044	0.011	0.001			
	3	0.010	0.057	0.201	0.267	0.215	0.117	0.042	0.009	0.001		
	4	0.001	0.011	0.088	0.200	0.251	0.205	0.111	0.037	0.006		
	5		0.001	0.026	0.103	0.201	0.246	0.201	0.103	0.026	0.001	
	6			0.006	0.037	0.111	0.205	0.251	0.200	0.088	0.011	0.001
	7			0.001	0.009	0.042	0.117	0.215	0.267	0.201	0.057	0.010
	8				0.001	0.011	0.044	0.121	0.233	0.302	0.194	0.075
	9					0.002	0.010	0.040	0.121	0.268	0.387	0.315
	10						0.001	0.006	0.028	0.107	0.349	0.599

TABLE III

BINOMIAL PROBABILITIES (*Continued*)

		p										
n	x	0.05	0.1	0.2	0.3	0.4	0.5	0.6	0.7	0.8	0.9	0.95
11	0	0.569	0.314	0.086	0.020	0.004						
	1	0.329	0.384	0.236	0.093	0.027	0.005	0.001				
	2	0.087	0.213	0.295	0.200	0.089	0.027	0.005	0.001			
	3	0.014	0.071	0.221	0.257	0.177	0.081	0.023	0.004			
	4	0.001	0.016	0.111	0.220	0.236	0.161	0.070	0.017	0.002		
	5		0.002	0.039	0.132	0.221	0.226	0.147	0.057	0.010		
	6			0.010	0.057	0.147	0.226	0.221	0.132	0.039	0.002	
	7			0.002	0.017	0.070	0.161	0.236	0.220	0.111	0.016	0.001
	8				0.004	0.023	0.081	0.177	0.257	0.221	0.071	0.014
	9				0.001	0.005	0.027	0.089	0.200	0.295	0.213	0.087
	10					0.001	0.005	0.027	0.093	0.236	0.384	0.329
	11							0.004	0.020	0.086	0.314	0.569
12	0	0.540	0.282	0.069	0.014	0.002						
	1	0.341	0.377	0.206	0.071	0.017	0.003					
	2	0.099	0.230	0.283	0.168	0.064	0.016	0.002				
	3	0.017	0.085	0.236	0.240	0.142	0.054	0.012	0.001			
	4	0.002	0.021	0.133	0.231	0.213	0.121	0.042	0.008	0.001		
	5		0.004	0.053	0.158	0.227	0.193	0.101	0.029	0.003		
	6			0.016	0.079	0.177	0.226	0.177	0.079	0.016		
	7			0.003	0.029	0.101	0.193	0.227	0.158	0.053	0.004	
	8			0.001	0.008	0.042	0.121	0.213	0.231	0.133	0.021	0.002
	9				0.001	0.012	0.054	0.142	0.240	0.236	0.085	0.017
	10					0.002	0.016	0.064	0.168	0.283	0.230	0.099
	11						0.003	0.017	0.071	0.206	0.377	0.341
	12							0.002	0.014	0.069	0.282	0.540
13	0	0.513	0.254	0.055	0.010	0.001						
	1	0.351	0.367	0.179	0.054	0.011	0.002					
	2	0.111	0.245	0.268	0.139	0.045	0.010	0.001				
	3	0.021	0.100	0.246	0.218	0.111	0.035	0.006	0.001			
	4	0.003	0.028	0.154	0.234	0.184	0.087	0.024	0.003			
	5		0.006	0.069	0.180	0.221	0.157	0.066	0.014	0.001		
	6		0.001	0.023	0.103	0.197	0.209	0.131	0.044	0.006		
	7			0.006	0.044	0.131	0.209	0.197	0.103	0.023	0.001	

TABLE III

BINOMIAL PROBABILITIES (*Continued*)

		p										
n	*x*	0.05	0.1	0.2	0.3	0.4	0.5	0.6	0.7	0.8	0.9	0.95
13	8			0.001	0.014	0.066	0.157	0.221	0.180	0.069	0.006	
	9				0.003	0.024	0.087	0.184	0.234	0.154	0.028	0.003
	10				0.001	0.006	0.035	0.111	0.218	0.246	0.100	0.021
	11					0.001	0.010	0.045	0.139	0.268	0.245	0.111
	12						0.002	0.011	0.054	0.179	0.367	0.351
	13							0.001	0.010	0.055	0.254	0.513
14	0	0.488	0.229	0.044	0.007	0.001						
	1	0.359	0.356	0.154	0.041	0.007	0.001					
	2	0.123	0.257	0.250	0.113	0.032	0.006	0.001				
	3	0.026	0.114	0.250	0.194	0.085	0.022	0.003				
	4	0.004	0.035	0.172	0.229	0.155	0.061	0.014	0.001			
	5		0.008	0.086	0.196	0.207	0.122	0.041	0.007			
	6		0.001	0.032	0.126	0.207	0.183	0.092	0.023	0.002		
	7			0.009	0.062	0.157	0.209	0.157	0.062	0.009		
	8			0.002	0.023	0.092	0.183	0.207	0.126	0.032	0.001	
	9				0.007	0.041	0.122	0.207	0.196	0.086	0.008	
	10				0.001	0.014	0.061	0.155	0.229	0.172	0.035	0.004
	11					0.003	0.022	0.085	0.194	0.250	0.114	0.026
	12					0.001	0.006	0.032	0.113	0.250	0.257	0.123
	13						0.001	0.007	0.041	0.154	0.356	0.359
	14							0.001	0.007	0.044	0.229	0.488
15	0	0.463	0.206	0.035	0.005							
	1	0.366	0.343	0.132	0.031	0.005						
	2	0.135	0.267	0.231	0.092	0.022	0.003					
	3	0.031	0.129	0.250	0.170	0.063	0.014	0.002				
	4	0.005	0.043	0.188	0.219	0.127	0.042	0.007	0.001			
	5	0.001	0.010	0.103	0.206	0.186	0.092	0.024	0.003			
	6		0.002	0.043	0.147	0.207	0.153	0.061	0.012	0.001		
	7			0.014	0.081	0.177	0.196	0.118	0.035	0.003		
	8			0.003	0.035	0.118	0.196	0.177	0.081	0.014		
	9			0.001	0.012	0.061	0.153	0.207	0.147	0.043	0.002	
	10				0.003	0.024	0.092	0.186	0.206	0.103	0.010	0.001
	11				0.001	0.007	0.042	0.127	0.219	0.188	0.043	0.005
	12					0.002	0.014	0.063	0.170	0.250	0.129	0.031
	13						0.003	0.022	0.092	0.231	0.267	0.135
	14							0.005	0.031	0.132	0.343	0.366
	15								0.005	0.035	0.206	0.463

To find the square root of any positive number rounded to two digits, use the following rule to decide whether to take the entry of the $\sqrt{n}$ or the $\sqrt{10n}$ column:

> **Move the decimal point an even number of places to the right or to the left until a number greater than or equal to 1 but less than 100 is reached. If the resulting number is less than 10 go to the $\sqrt{n}$ column; if it is 10 or more go to the $\sqrt{10n}$ column.**

Thus, to find the square root of 62,000 or 0.017 we go to the $\sqrt{n}$ column since the decimal point has to be moved, respectively, four places to the left to give 6.2 or two places to the right to give 1.7. Similarly, to find the square root of 2,100 or 0.000033 we go to the $\sqrt{10n}$ column since the decimal point has to be moved, respectively, two places to the left to give 21 or six places to the right to give 33.

Having found the entry in the appropriate column of Table IV, the only thing that remains to be done is to put the decimal point in the right place in the result. To this end we use the following rule:

> **Having previously moved the decimal point an even number of places to the left or to the right to get a number greater than or equal to 1 but less than 100, the decimal point of the appropriate entry of Table IV is moved half as many places in the opposite direction.**

For example, to find the square root of 62,000 we first note that the decimal point has to be moved *four places to the left* to give 6.2. We thus take the entry of the $\sqrt{n}$ column corresponding to 6.2, move its decimal point *two places to the right,* and get $\sqrt{62{,}000} = 248.998$. Similarly, to find the square root of 0.000033 we note that the decimal point has to be moved *six places to the right* to give 33. We thus take the entry of the $\sqrt{10n}$ column corresponding to 33, move the decimal point *three places to the left,* and get $\sqrt{0.000033} = 0.00574456$.

TABLE IV

SQUARE ROOTS

n	$\sqrt{n}$	$\sqrt{10n}$	n	$\sqrt{n}$	$\sqrt{10n}$
1.0	1.00000	3.16228	**3.5**	1.87083	5.91608
1.1	1.04881	3.31662	**3.6**	1.89737	6.00000
1.2	1.09545	3.46410	**3.7**	1.92354	6.08276
1.3	1.14018	3.60555	**3.8**	1.94936	6.16441
1.4	1.18322	3.74166	**3.9**	1.97484	6.24500
1.5	1.22474	3.87298	**4.0**	2.00000	6.32456
1.6	1.26491	4.00000	**4.1**	2.02485	6.40312
1.7	1.30384	4.12311	**4.2**	2.04939	6.48074
1.8	1.34164	4.24264	**4.3**	2.07364	6.55744
1.9	1.37840	4.35890	**4.4**	2.09762	6.63325
2.0	1.41421	4.47214	**4.5**	2.12132	6.70820
2.1	1.44914	4.58258	**4.6**	2.14476	6.78233
2.2	1.48324	4.69042	**4.7**	2.16795	6.85565
2.3	1.51658	4.79583	**4.8**	2.19089	6.92820
2.4	1.54919	4.89898	**4.9**	2.21359	7.00000
2.5	1.58114	5.00000	**5.0**	2.23607	7.07107
2.6	1.61245	5.09902	**5.1**	2.25832	7.14143
2.7	1.64317	5.19615	**5.2**	2.28035	7.21110
2.8	1.67332	5.29150	**5.3**	2.30217	7.28011
2.9	1.70294	5.38516	**5.4**	2.32379	7.34847
3.0	1.73205	5.47723	**5.5**	2.34521	7.41620
3.1	1.76068	5.56776	**5.6**	2.36643	7.48331
3.2	1.78885	5.65685	**5.7**	2.38747	7.54983
3.3	1.81659	5.74456	**5.8**	2.40832	7.61577
3.4	1.84391	5.83095	**5.9**	2.42899	7.68115

TABLE IV

SQUARE ROOTS (*Continued*)

n	$\sqrt{n}$	$\sqrt{10n}$	n	$\sqrt{n}$	$\sqrt{10n}$
6.0	2.44949	7.74597	**8.0**	2.82843	8.94427
6.1	2.46982	7.81025	**8.1**	2.84605	9.00000
6.2	2.48998	7.87401	**8.2**	2.86356	9.05539
6.3	2.50998	7.93725	**8.3**	2.88097	9.11043
6.4	2.52982	8.00000	**8.4**	2.89828	9.16515
6.5	2.54951	8.06226	**8.5**	2.91548	9.21954
6.6	2.56905	8.12404	**8.6**	2.93258	9.27362
6.7	2.58844	8.18535	**8.7**	2.94958	9.32738
6.8	2.60768	8.24621	**8.8**	2.96648	9.38083
6.9	2.62679	8.30662	**8.9**	2.98329	9.43398
7.0	2.64575	8.36660	**9.0**	3.00000	9.48683
7.1	2.66458	8.42615	**9.1**	3.01662	9.53939
7.2	2.68328	8.48528	**9.2**	3.03315	9.59166
7.3	2.70185	8.54400	**9.3**	3.04959	9.64365
7.4	2.72029	8.60233	**9.4**	3.06594	9.69536
7.5	2.73861	8.66025	**9.5**	3.08221	9.74679
7.6	2.75681	8.71780	**9.6**	3.09839	9.79796
7.7	2.77489	8.77496	**9.7**	3.11448	9.84886
7.8	2.79285	8.83176	**9.8**	3.13050	9.89949
7.9	2.81069	8.88819	**9.9**	3.14643	9.94987

ANSWERS TO ODD-NUMBERED EXERCISES

Page 10

1. (a) 2; (b) 2.
3. (b) 40; (c) 60; (d) 30.
5. There are, respectively, 6 ways and 4 ways, but we *cannot* conclude that he is more likely to distribute two of the three pieces of candy unless we assume, for example, that all the 20 possibilities are equally likely.
7. In 21 ways.
9. 32 and 10.
11. (a) 4; (b) 16; (c) 12.
13. (a) 6; (b) 18.
15. 40.
17. (a) 72; (b) 6; (c) 66.
19. (a) 32,768; (b) 245,760.
21. 1,023.

Page 20

1. 336.
3. 60,480.
5. 362,880 and $\frac{1}{9}$.
7. (a) 40,320; (b) 24; (c) 384; (d) 1,152; (e) 1,152.
9. $(n-1)!$; (a) 5,040; (b) 2,520.
11. (a) 6,720; (b) 180; (c) 1,680; (d) 1,120; (e) 3,360.
13. (a) 40,320; (b) 70; (c) 420.
15. (a) true; (b) false; (c) true; (d) false.

Page 29

1. 55.
3. 924.

5. 11,628 and 73.7%.
7. (a) 455; (b) 5,005; (c) 455.
9. 120.
11. 207,900.
13. (a) 120; (b) 10; (c) 90; (d) 45.5%; (e) 3.55%.
15. (a) 6; (b) 30; (c) 20; (d) $6+30+20 = 56$.
17. 180.
19. 5,040.
21. (a) $56 = 35+21$, checks; (b) $220 = 4 \cdot 55$, checks; (c) $792 = 462+330$, checks; (d) $7 \cdot 120 = 10 \cdot 84$, checks.
23. $5+10+10+5+1 = 31$.

Page 43

1. (a) $\frac{1}{3}$; (b) $\frac{1}{2}$; (c) $\frac{1}{3}$.
3. $\frac{1}{4}$, $\frac{1}{2}$, and $\frac{1}{4}$.
5. (a) $\frac{1}{2}$; (b) $\frac{11}{25}$; (c) $\frac{3}{50}$; (d) $\frac{37}{50}$.
7. $\frac{1}{2}$.
9. $\frac{2}{3}$.
13. (a) $\frac{12}{19}$; (b) $\frac{28}{57}$.
15. $\frac{3}{14}$.
17. (a) $\frac{91}{300}$; (b) $\frac{11}{60}$; (c) $\frac{77}{150}$; no coincidence.
17. (a) $\dfrac{1}{32{,}768}$; (b) $\dfrac{3{,}003}{32{,}768}$; (c) $\dfrac{6{,}435}{32{,}768}$; (d) $\dfrac{1}{32{,}768}$.
21. (a) $\frac{5}{14}$; (b) $\frac{3}{28}$; (c) $\frac{15}{28}$; no coincidence.
23. $\frac{1}{10}$.
25. (a) 784 to 4,061; (b) 4,844 to 1; (c) $\dfrac{4{,}350}{4{,}845}$.
27. (a) 1 to 19; (b) 1 to 3; (c) 11 to 2.
29. (a) $\frac{13}{15}$; (b) $\frac{1}{4}$ and $\frac{5}{8}$; (c) $\frac{5}{6}$.
31. The ratio of 1 to 0 is undefined (meaningless).
33. Most persons like some cheeses more than others.

Page 54

1. 0.76.
3. 0.18.
5. (a) 0.40; (b) 2 to 3; (c) 3 to 2; (d) it would favor us, as he should bet only 15 cents against our dime.

7. (a) 0.75; (b) 3 to 1; (c) 1 to 3; (d) we would be favored, as we should bet \$12 against his \$4 to make it fair.
9. $\frac{5}{8}$.
11. $\frac{5}{16}$.
13. At least $\frac{7}{9}$.
15. $\frac{5}{6} \leqq p < \frac{6}{7}$.
17. $\frac{3}{4} \leqq p < \frac{4}{5}$.
19. When we have more information, we can be more certain that the value of a probability is correct, or close, but the probability, itself, can be large or small.
21. (a) We must refer to the reliability, or truthfulness, of similar witnesses in the past; (b) We might ask ourselves at what odds we would be willing to bet that the testimony is true.
25. Draw a tree diagram; it should show that in two of the three possibilities the other coin is also a penny.

Page 64

1. \$0.76.
3. The expectations are, respectively, \$0.25 and \$0.26, so that it is preferable to make a drawing from Box B.
5. \$0.25, so that it is not worthwhile to spend \$0.30 on gasoline.
7. (a) $2\frac{1}{2}$ cents, and it is not worthwhile to spend the 8 cents on postage.
9. \$1,600, and the player would be better off to split the prize money.
11. \$2.00.
13. \$5.20.
15. The expectation is 10 cents, so that it is worthwhile to spend 8 cents for a package, 10 cents is a fair price, and it is not worthwhile to spend 12 cents.
17. 212.22 guests.
19. \$430.00.
21. 1.32 purchases.

Page 71

3. 0.16.
5. (a) $p < \frac{1}{3}$; (b) $p > \frac{1}{3}$; (c) $p = \frac{1}{3}$.
7. (a) the probability is less than $\frac{3}{5}$; (b) the probability is greater than $\frac{3}{5}$; (c) the probability equals $\frac{3}{5}$.
9. \$16.50.
11. (a) $U < \$141$; (b) $\$101 \leqq U < \$131$; (c) $U = \$125$.

13. −5 utiles.
15. 1.56 utiles.

Page 80

1. (a)

	Franchise	*No franchise*
Build arena	2,050,000	−500,000
Do not build	1,000,000	100,000

(b) since \$400,000 exceeds \$350,000, it is better not to build the arena; (c) since \$520,000 exceeds \$460,000, it is better to build the arena; (d) he would be against building the new arena because of the possible loss of \$500,000; (e) he would favor building the new arena because of the possible profit of \$2,050,000; (f)

	Franchise	*No franchise*
Build arena	0	600,000
Do not build	1,050,000	0

and the greatest possible opportunity loss is least when they build; (g) since expected opportunity losses of \$350,000 are preferable to expected opportunity losses of \$400,000, it would be better not to build the arena.

3. (a) Build the arena; (b) not build the arena.
5. (a) Since \$133,333 exceeds −\$171,666, it is better to put the new tire on the market; (b) since −\$118,000 exceeds −\$240,000, it is better not to put the new tire on the market; (c) in either case there is an expected loss of \$133,333, so it does not matter whether the tire is put on the market.
7. (a) 11.5 miles; (b) 13.5 miles.
9. (a) Since $\frac{20}{7}$ is less than $\frac{26}{7}$, it would be better not to take the coat; (b) since 2 is less than 5, it would be better to take the coat; (c) since the expected inconvenience is the same, it would not matter whether she takes the coat; (d) since 6 is less than 30, she should take her coat.
11. (a) $3\frac{1}{3}$; (b) $3\frac{1}{3}$; this is preferable to the gamble of Exercise 10 because the maximum expected inconvenience is less.
13. (a) \$750,000, which makes it worthwhile to spend the \$50,000; (b) $10\frac{1}{3}$ miles, which cost 31 cents, and hence it is not worthwhile to spend the 10 cents.
15. (a) Send roses; (b) send roses.

Page 94

1. (d) (1, 0) and (0, 1); (e) (1, 1, 0), (1, 0, 1), and (0, 1, 1); (f) (1, 1, 0), (1, 0, 1), (0, 1, 1), and (1, 1, 1); (g) answers to first and fourth questions are right, and answers to second and third questions are wrong; (h) (1, 1, 0, 0), (1, 0, 1, 0), (1, 0, 0, 1), (0, 1, 1, 0), (0, 1, 0, 1), and (0, 0, 1, 1); (i) answers to the first three questions are wrong, and the answers to the fourth and fifth questions are right; (j) (1, 1, 1, 0, 0), (1, 1, 1, 1, 0), (1, 1, 1, 0, 1), and (1, 1, 1, 1, 1); (k) 2, 3, and 4.
3. (a) (2, 2); (b) (3, 3); (c) (1, 2) and (2, 1).
5. (a) (0, 0), (1, 1), and (2, 2); (b) (2, 0), (2, 1), and (2, 2); (c) (1, 0) and (2, 1); (d) (0, 0), (1, 0), (0, 1), (2, 0), (1, 1), and (0, 2).
7. (a) Mutually exclusive; (b) mutually exclusive; (c) not mutually exclusive.
9. (a) Exactly one cab is out on a call; (b) at least two of the cabs are not out on a call; (c) at most one of the cabs is operative.
11. (a) (1, 4), (2, 3), (3, 2), (2, 4), (3, 3), and (3, 4); (b) (0, 0), (1, 1), (2, 2), and (3, 3); (c) (1, 0), (2, 0), (2, 1), (3, 0), (3, 1), and (3, 2).
13. (a) Not mutually exclusive; (b) mutually exclusive; (c) not mutually exclusive; (d) not mutually exclusive; (e) mutually exclusive; (f) not mutually exclusive.
15. (a) {Car 5, Car 6, Car 7, Car 8}; (b) {Car 2, Car 4, Car 5, Car 7}; (c) {Car 1, Car 8}; (d) {Car 3, Car 4, Car 7, Car 8}.

Page 102

1. (a) A saleslady working at the department store is not satisfied with working conditions; (b) a saleslady working at the department store is not satisfied with her wages; (c) a saleslady working at the department store is satisfied with the working conditions and/or her wages; (d) a saleslady working at the department store is satisfied with neither the working conditions nor her wages; (e) a saleslady working at the department store is not satisfied with her wages and/or the working conditions; (f) a saleslady working at the department store is satisfied with her wages but not with the working conditions.
3. (a) Los Angeles, Ontario, Pasadena, Riverside, Santa Monica, and Burbank; (b) San Diego, Ontario, Pasadena, Riverside, Santa Monica, and Burbank; (c) Santa Barbara; (d) $\emptyset$; (e) San Diego, Ontario, Santa Barbara, Riverside; (f) San Diego, Ontario, and Santa Barbara; (g) all except Los Angeles; (h) Los Angeles, Pasadena, Santa Monica, and Burbank; (i) Los Angeles, Pasadena, Santa Monica, and Burbank.
5. (a) (0, 0), (0, 1), (0, 2), (0, 3), (0, 4), (1, 1), (1, 2), (1, 3), and (2, 2); (b) (0, 0), (0, 1), (0, 2), (0, 3), (0, 4), (1, 0), (2, 0), (3, 0), and (4, 0); (c) (0, 4), (1, 3), (2, 2),

(3, 1), (4, 0), (0, 1), and (1, 0); (d) (2, 1) and (3, 1); (e) (0, 0), (0, 1), (0, 2), (0, 3), (0, 4), (2, 0), (2, 1), (2, 2), (3, 0), (3, 1), and (4, 0); (f) (1, 3), (2, 2), and (3, 1); (g) (0, 3); (h) the entire set; (i) (0, 3), (0, 4), and (1, 3).

7. (a) (0, 0, 0), (1, 0, 0), (0, 1, 0), (0, 0, 1), (1, 1, 0), (1, 0, 1), (0, 1, 1), and (1, 1, 1); (b) $F = \{(1, 1, 0), (1, 1, 1)\}$, $G = \{(0, 0, 0), (1, 0, 0), (0, 1, 0), (0, 0, 1), (1, 1, 0), (1, 0, 1), (0, 1, 1)\}$, $H = \{(0, 0, 0), (1, 1, 1)\}$, and $I = \{(0, 1, 0), (1, 1, 0), (0, 1, 1), (1, 1, 1)\}$; (c) G' is the event of getting three heads, I' is the event that the second flip comes up tails, $H \cup F$ is the event that either the first two flips come up heads or all three come up tails, $G \cap I$ is the event that the second flip comes up heads but at least one of the other flips comes up tails, $F \cap G'$ is the event that all three flips come up heads, and $H' \cap I'$ is the event that the second flip comes up tails and at least one flip comes up heads; (d) $G' = \{(1, 1, 1)\}$, $I' = \{(0, 0, 0), (1, 0, 0), (0, 0, 1), (1, 0, 1)\}$, $H \cup F = \{(1, 1, 1), (0, 0, 0), (1, 1, 0)\}$, $G \cap I = \{(0, 1, 0), (1, 1, 0), (0, 1, 1)\}$, $F \cap G' = \{(1, 1, 1)\}$, and $H' \cap I' = \{(1, 0, 1), (1, 0, 0), (0, 0, 1)\}$.

9. (a) Fewer than two cabs are out on a call; (b) either 0 or 1 operative cabs are not out on a call; (c) unless only one cab is operative, at least one of the operative cabs is not out on a call; (d) one cab is operative; (e) at least one cab is out on a call; (f) one cab is operative and it is not out on a call; (g) at least two operative cabs are not out on a call; (h) at least one cab is operative; (i) same number of cabs are operative as out on a call.

11. (a) The number of tables used in the larger dining room does not exceed the number of tables used in the smaller dining room exactly by two; (b) altogether fewer than five tables are being used; (c) the number of tables used in the large dining room does not exceed the number of tables used in the small dining room; (d) none of the tables in the small dining room and all of the tables in the large dining room are being used; (e) four of the tables in the large dining room and at least one of the tables in the small dining room are being used; (f) either one dining room is empty or altogether four tables are being used; (g) anything but two tables being used in each dining room; (h) none of the dining rooms is empty and altogether fewer than five tables are being used; (i) all four tables in the large dining room are being used.

15. (a) 4; (b) 1 and 2; (c) 4 and 6; (d) 1, 2, 3, 4, 5, 7; (e) 7 and 8.

17. (a) 7; (b) 1 and 2; (c) 3 and 6; (d) 8; (e) 1, 3, 4, and 6.

19. (a) 3, 5, 6, and 8; (b) 1, 2, 3, 4, 5, and 7; (c) 1 and 4; (d) 1, 2, 3, 5, 7, and 8; (e) 6 and 8; (f) 4.

Page 109

1. $\frac{1}{4}$.

3. The total for the four regions is 430, which exceeds 400, and the survey should be questioned.
5. (a) $\frac{4}{15}$; (b) $\frac{2}{5}$; (c) $\frac{31}{60}$.

Page 119

1. The first claim violates Postulate 2 because the probabilities do not add up to one; the second claim violates Postulate 1 because one of the probabilities is negative; the third claim violates Postulate 2 since the sum of the probabilities exceeds one.
3. The corresponding probabilities are $\frac{3}{5}$, $\frac{1}{5}$, and $\frac{1}{10}$, and since their sum is not one, this violates Postulate 2.
5. The respective probabilities are $\frac{1}{3}$, $\frac{1}{6}$, and $\frac{1}{2}$, and since $\frac{1}{3}+\frac{1}{6}=\frac{1}{2}$ they are consistent.
7. The respective probabilities are $\frac{1}{3}$ and $\frac{1}{4}$, and since $\frac{1}{3}+\frac{1}{4}=\frac{7}{12}$ exceeds $\frac{1}{2}$, the claims are consistent.
9. The respective probabilities are $\frac{2}{3}$ and $\frac{5}{6}$, and since $\frac{2}{3}+\frac{5}{6}=\frac{9}{6}$ exceeds one, Postulate 2 is violated.
11. (a) 0.77; (b) 0.62; (c) 0.62.
13. (a) 0.25; (b) 0.75; (c) 0.99.
15. (a) 0.31; (b) 0.14; (c) 0.33; (d) 0.23; (e) 0.54; (f) 0.32.
17. (a) 0.62; (b) 0.66; (c) 0.33; (d) 0.34.
19. (a) $\frac{1}{3}$, $\frac{2}{5}$, $\frac{1}{5}$, and $\frac{2}{15}$; (b) $\frac{1}{3}$, $\frac{2}{5}$, and $\frac{2}{5}$; (c) $\frac{3}{5}$, $\frac{3}{5}$, $\frac{7}{15}$, $\frac{2}{15}$, $\frac{11}{15}$, $\frac{1}{5}$, $\frac{1}{15}$, $\frac{4}{15}$, and $\frac{1}{5}$.
21. (a) $\frac{3}{10}$, $\frac{1}{5}$, and $\frac{3}{10}$; (b) $\frac{1}{5}$, $\frac{1}{5}$, $\frac{3}{20}$, and $\frac{2}{5}$; (c) $\frac{17}{20}$, $\frac{7}{10}$, $\frac{1}{2}$, $\frac{1}{20}$, $\frac{3}{20}$, $\frac{11}{20}$, $\frac{19}{20}$, $\frac{3}{10}$, and $\frac{1}{5}$.
23. (a) 0.44; (b) 0.46; (c) 0.46.
25. (a) 0.30, 0.40, and 0.50; (b) 0.09, 0.12, 0.35, 0.21, and 0.40; (c) 0.70, 0.58, 0.15, 0.70, 0.42, and 0.09.

Page 129

5. (a) 0.16+0.23 does not equal 0.38; (b) the first probability cannot exceed the second; (c) the two probabilities are 0.75 and 0.80, respectively, and the first should not be smaller than the second; (d) the sum of the probabilities exceeds one; (e) the second probability cannot exceed the first; (f) the sum of the probabilities is less than one.
7. (a) 0.33; (b) 0.77; (c) 0.11; (d) 0.55; (e) 0.78; (f) 0.22.
9. (a) 0.25; (b) 0.20; (c) 0.05; (d) 0.40.
11. (a) 0.13; (b) 0.98; (c) 0.11.
13. (a) $\frac{37}{45}$; (b) $\frac{19}{45}$; (c) $\frac{13}{45}$; (d) $\frac{43}{45}$; (e) $\frac{2}{45}$.
15. 0.70, 0.50, 0.40, and 0.80.
17. 0.42, 0.40, 0.38, 0.12, 0.15, 0.20, 0.06, and 0.79.
19. 0.98.

Page 144

1. The probabilities cannot be added because they pertain to different sample spaces.
3. (a) $P(W' \mid I)$; (b) $P(I' \mid W)$; (c) $P(I' \mid W')$; (d) $P(W \mid I')$.
5. (a) The probability that an astronaut who is a well trained scientist is a member of the armed forces; (b) the probability that an astronaut who is not a member of the armed services is a well trained scientist; (c) the probability that an astronaut who was not a test pilot is not a member of the armed services; (d) the probability that an astronaut who was a test pilot is a well trained scientist but not a member of the armed services; (e) the probability that an astronaut who is a well trained scientist and was a test pilot is a member of the armed services; (f) the probability that an astronaut who is a well trained scientist and/or was a test pilot is a member of the armed services.
7. (a) $\frac{3}{5}$; (b) $\frac{2}{5}$; (c) $\frac{3}{10}$; (d) $\frac{7}{10}$; (e) $\frac{1}{5}$; (f) $\frac{2}{3}$; (g) $\frac{1}{3}$; (h) $\frac{3}{7}$; (i) $\frac{3}{4}$; (j) $\frac{3}{10}$.
9. (a) $\frac{5}{13}$; (b) $\frac{7}{13}$; (c) $\frac{3}{13}$; (d) $\frac{2}{13}$; (e) $\frac{4}{13}$; (f) $\frac{3}{7}$; (g) $\frac{1}{3}$; (h) $\frac{3}{5}$; (i) $\frac{1}{2}$.
11. (a) $\frac{7}{10}$; (b) $\frac{79}{200}$; (c) $\frac{63}{200}$; (d) $\frac{2}{25}$; (e) $\frac{77}{200}$; (f) $\frac{63}{79}$; (g) $\frac{9}{20}$; (h) $\frac{77}{121}$; (i) $\frac{4}{15}$.
13. (a) $\frac{7}{9}$; (b) $\frac{14}{23}$; (c) $\frac{9}{82}$; (d) $\frac{73}{77}$.
15. (a) 0.50; (b) 0.40; (c) 0.50.
17. $\frac{44}{105}$.
19. (a) 0.90; (b) 0.96.
21. (a) $\dfrac{0.24}{0.40} = 0.60$, checks; (b) $\dfrac{0.24}{0.60} = 0.40$, checks; (c) $\dfrac{0.36}{0.60} = 0.60$, checks; $\dfrac{0.16}{0.40} = 0.40$, checks.
25. (a) Not independent; (b) not independent.
27. (a) Dependent; (b) independent; (c) independent; (d) dependent; (e) dependent; (f) independent; (g) dependent; (h) dependent; (i) dependent.
29. (a) $\frac{25}{102}$; (b) $\frac{1}{4}$.
31. (a) 0.343; (b) 0.113; (c) 0.262.
33. 0.05.
35. (a) 0.048; (b) 0.00768.
37. $P(A \mid B) = P(A) = \frac{1}{2}$, $P(A \mid C) = P(A) = \frac{1}{2}$, but $P(A \mid B \cap C) = 1 \neq P(A)$.

Page 162

1. 0.573.
3. 0.876.
7. 0.428.

9. (a) 0.24; (b) 0.010.
13. 0.462.
15. 71 to 29.
19. 0.652.
21. $\frac{3}{5}$.
23. (a) $x_1+x_2+x_3+x_4+x_5+x_6$; (b) $P(A_1)+P(A_2)+P(A_3)+P(A_4)$;
(c) $a_1p_1+a_2p_2+\ldots+a_np_n$;
(d) $P(B_1|A)+P(B_2|A)+P(B_3|A)+P(B_4|A)+P(B_5|A)$;
(e) $P(A\cap B_1)+P(A\cap B_2)+P(A\cap B_3)$; (f) $(1-p_2)+(1-p_3)+(1-p_4)$.

Page 172

3. (a) the probabilities that 0, 1, 2, or 3 tables are used are, respectively, 0.25, 0.25, 0.25, and 0.25; (b) the probabilities that 0, 1, 2, 3, or 4 tables are used are, respectively, 0.20, 0.20, 0.20, 0.20, and 0.20; (c) the probabilities that 0, 1, 2, 3, 4, 5, 6, or 7 tables are used are, respectively, 0.05, 0.10, 0.15, 0.20, 0.20, 0.15, 0.10, and 0.05.
7. (a) No, sum of the probabilities exceeds one; (b) yes; (c) yes; (d) no, sum exceeds 1.
9. (a) 0.60; (b) 0.40.
11. 1.25.
13. (a) 1.5; (b) 2; (c) 3.5.

Page 183

1. (a) $\frac{625}{3{,}888}$; (b) $\frac{7{,}500}{7{,}776}$; (c) $\frac{7}{64}$; (d) $\frac{37}{256}$.
3. (a) $\frac{729}{4{,}096}$; (b) $\frac{1{,}458}{4{,}096}$; (c) $\frac{1{,}215}{4{,}096}$; (d) $\frac{694}{4{,}096}$.
5. (a) 0.324; (b) 0.324.
7. $\frac{37}{256}$.
9. (a) $\frac{1}{256}$; (b) $\frac{81}{256}$; (c) $\frac{12}{256}$; (d) $\frac{162}{256}$.
11. (a) 0.334; (b) 0.438; (c) 0.902.
13. (a) 0.061; (b) 0.000; (c) 0.982; (d) 0.584.
15. (a) 0.308; (b) 0.548.
17. (a) 0.460; (b) 0.069.
19. (a) 0.123; (b) 0.832; (c) 0.352.
21. 10.792.
23. (a) 0.698; (b) 0.759; (c) 0.173.
25. (a) 0.667, 0.222, and 0.111; (b) 0.717, 0.217, and 0.066.

Page 197

1. 0.36.
3. 0.75.
5. (a) 0.071; (b) 0.381; (c) 0.429; (d) 0.114; (e) 0.05.
7. (a) 0.524; (b) 0.419; (c) 0.057.
9. (a) 0.288; (b) 0.167.
11. 0.336, error is 0.007.
13. 0.059.
15. 0.015.
17. (a) 0.073; (b) 0.053; (c) 0.349.
19. 0.0114.
21. 0.0367.
23. 0.20.
25. 0.0945.
27. 0.021.
29. (a) $\dfrac{\binom{a_1}{x_1}\binom{a_2}{x_2}\cdot\ldots\cdot\binom{a_k}{x_k}}{\binom{N}{n}}$; (b) 0.195.

Page 208

1. 1.25.
3. 1.73.
7. (a) 1.15; (b) 1.73: (c) 1.22; (d) 0.65.
9. (a) 0.69; (b) 0.69.
11. (a) 1.00; (b) 1.05.
13. (a) 0.6; (b) 0.6.
15. 0.80.
17. (a) 0.6; (b) 0.71; (c) 0.80; (d) 0.6.
19. 1.41.

Page 215

1. (a) 32 and 4; (b) the probability is at least 0.889 that he will get between 20 and 44 correct answers; (c) 0.938.
3. (a) The probability is at least 0.938 that between 90 and 158 marriage licenses will be issued; (b) 0.028.
5. The probability is at least 0.908 that there will be at most 5 robberies.
7. The probability is at least 0.96.

INDEX

A CATALOG OF SELECTED

DOVER BOOKS

IN SCIENCE AND MATHEMATICS

Astronomy

BURNHAM'S CELESTIAL HANDBOOK, Robert Burnham, Jr. Thorough guide to the stars beyond our solar system. Exhaustive treatment. Alphabetical by constellation: Andromeda to Cetus in Vol. 1; Chamaeleon to Orion in Vol. 2; and Pavo to Vulpecula in Vol. 3. Hundreds of illustrations. Index in Vol. 3. 2,000pp. 6⅛ x 9¼.
Vol. I: 23567-X
Vol. II: 23568-8
Vol. III: 23673-0

EXPLORING THE MOON THROUGH BINOCULARS AND SMALL TELESCOPES, Ernest H. Cherrington, Jr. Informative, profusely illustrated guide to locating and identifying craters, rills, seas, mountains, other lunar features. Newly revised and updated with special section of new photos. Over 100 photos and diagrams. 240pp. 8¼ x 11. 24491-1

THE EXTRATERRESTRIAL LIFE DEBATE, 1750–1900, Michael J. Crowe. First detailed, scholarly study in English of the many ideas that developed from 1750 to 1900 regarding the existence of intelligent extraterrestrial life. Examines ideas of Kant, Herschel, Voltaire, Percival Lowell, many other scientists and thinkers. 16 illustrations. 704pp. 5⅜ x 8½. 40675-X

THEORIES OF THE WORLD FROM ANTIQUITY TO THE COPERNICAN REVOLUTION, Michael J. Crowe. Newly revised edition of an accessible, enlightening book recreates the change from an earth-centered to a sun-centered conception of the solar system. 242pp. 5⅜ x 8½. 41444-2

A HISTORY OF ASTRONOMY, A. Pannekoek. Well-balanced, carefully reasoned study covers such topics as Ptolemaic theory, work of Copernicus, Kepler, Newton, Eddington's work on stars, much more. Illustrated. References. 521pp. 5⅜ x 8½. 65994-1

A COMPLETE MANUAL OF AMATEUR ASTRONOMY: Tools and Techniques for Astronomical Observations, P. Clay Sherrod with Thomas L. Koed. Concise, highly readable book discusses: selecting, setting up and maintaining a telescope; amateur studies of the sun; lunar topography and occultations; observations of Mars, Jupiter, Saturn, the minor planets and the stars; an introduction to photoelectric photometry; more. 1981 ed. 124 figures. 26 halftones. 37 tables. 335pp. 6½ x 9¼. 42820-6

AMATEUR ASTRONOMER'S HANDBOOK, J. B. Sidgwick. Timeless, comprehensive coverage of telescopes, mirrors, lenses, mountings, telescope drives, micrometers, spectroscopes, more. 189 illustrations. 576pp. 5⅝ x 8¼. (Available in U.S. only.) 24034-7

STARS AND RELATIVITY, Ya. B. Zel'dovich and I. D. Novikov. Vol. 1 of *Relativistic Astrophysics* by famed Russian scientists. General relativity, properties of matter under astrophysical conditions, stars, and stellar systems. Deep physical insights, clear presentation. 1971 edition. References. 544pp. 5⅝ x 8¼. 69424-0

Chemistry

THE SCEPTICAL CHYMIST: The Classic 1661 Text, Robert Boyle. Boyle defines the term "element," asserting that all natural phenomena can be explained by the motion and organization of primary particles. 1911 ed. viii+232pp. 5⅜ x 8½. 42825-7

RADIOACTIVE SUBSTANCES, Marie Curie. Here is the celebrated scientist's doctoral thesis, the prelude to her receipt of the 1903 Nobel Prize. Curie discusses establishing atomic character of radioactivity found in compounds of uranium and thorium; extraction from pitchblende of polonium and radium; isolation of pure radium chloride; determination of atomic weight of radium; plus electric, photographic, luminous, heat, color effects of radioactivity. ii+94pp. 5⅜ x 8½. 42550-9

CHEMICAL MAGIC, Leonard A. Ford. Second Edition, Revised by E. Winston Grundmeier. Over 100 unusual stunts demonstrating cold fire, dust explosions, much more. Text explains scientific principles and stresses safety precautions. 128pp. 5⅜ x 8½. 67628-5

THE DEVELOPMENT OF MODERN CHEMISTRY, Aaron J. Ihde. Authoritative history of chemistry from ancient Greek theory to 20th-century innovation. Covers major chemists and their discoveries. 209 illustrations. 14 tables. Bibliographies. Indices. Appendices. 851pp. 5⅜ x 8½. 64235-6

CATALYSIS IN CHEMISTRY AND ENZYMOLOGY, William P. Jencks. Exceptionally clear coverage of mechanisms for catalysis, forces in aqueous solution, carbonyl- and acyl-group reactions, practical kinetics, more. 864pp. 5⅜ x 8½. 65460-5

ELEMENTS OF CHEMISTRY, Antoine Lavoisier. Monumental classic by founder of modern chemistry in remarkable reprint of rare 1790 Kerr translation. A must for every student of chemistry or the history of science. 539pp. 5⅜ x 8½. 64624-6

THE HISTORICAL BACKGROUND OF CHEMISTRY, Henry M. Leicester. Evolution of ideas, not individual biography. Concentrates on formulation of a coherent set of chemical laws. 260pp. 5⅜ x 8½. 61053-5

A SHORT HISTORY OF CHEMISTRY, J. R. Partington. Classic exposition explores origins of chemistry, alchemy, early medical chemistry, nature of atmosphere, theory of valency, laws and structure of atomic theory, much more. 428pp. 5⅜ x 8½. (Available in U.S. only.) 65977-1

GENERAL CHEMISTRY, Linus Pauling. Revised 3rd edition of classic first-year text by Nobel laureate. Atomic and molecular structure, quantum mechanics, statistical mechanics, thermodynamics correlated with descriptive chemistry. Problems. 992pp. 5⅜ x 8½. 65622-5

FROM ALCHEMY TO CHEMISTRY, John Read. Broad, humanistic treatment focuses on great figures of chemistry and ideas that revolutionized the science. 50 illustrations. 240pp. 5⅜ x 8½. 28690-8

Engineering

DE RE METALLICA, Georgius Agricola. The famous Hoover translation of greatest treatise on technological chemistry, engineering, geology, mining of early modern times (1556). All 289 original woodcuts. 638pp. 6¾ x 11. 60006-8

FUNDAMENTALS OF ASTRODYNAMICS, Roger Bate et al. Modern approach developed by U.S. Air Force Academy. Designed as a first course. Problems, exercises. Numerous illustrations. 455pp. 5⅜ x 8½. 60061-0

DYNAMICS OF FLUIDS IN POROUS MEDIA, Jacob Bear. For advanced students of ground water hydrology, soil mechanics and physics, drainage and irrigation engineering, and more. 335 illustrations. Exercises, with answers. 784pp. 6⅛ x 9¼. 65675-6

THEORY OF VISCOELASTICITY (Second Edition), Richard M. Christensen. Complete, consistent description of the linear theory of the viscoelastic behavior of materials. Problem-solving techniques discussed. 1982 edition. 29 figures. xiv+364pp. 6⅛ x 9¼. 42880-X

MECHANICS, J. P. Den Hartog. A classic introductory text or refresher. Hundreds of applications and design problems illuminate fundamentals of trusses, loaded beams and cables, etc. 334 answered problems. 462pp. 5⅜ x 8½. 60754-2

MECHANICAL VIBRATIONS, J. P. Den Hartog. Classic textbook offers lucid explanations and illustrative models, applying theories of vibrations to a variety of practical industrial engineering problems. Numerous figures. 233 problems, solutions. Appendix. Index. Preface. 436pp. 5⅜ x 8½. 64785-4

STRENGTH OF MATERIALS, J. P. Den Hartog. Full, clear treatment of basic material (tension, torsion, bending, etc.) plus advanced material on engineering methods, applications. 350 answered problems. 323pp. 5⅜ x 8½. 60755-0

A HISTORY OF MECHANICS, René Dugas. Monumental study of mechanical principles from antiquity to quantum mechanics. Contributions of ancient Greeks, Galileo, Leonardo, Kepler, Lagrange, many others. 671pp. 5⅜ x 8½. 65632-2

STABILITY THEORY AND ITS APPLICATIONS TO STRUCTURAL MECHANICS, Clive L. Dym. Self-contained text focuses on Koiter postbuckling analyses, with mathematical notions of stability of motion. Basing minimum energy principles for static stability upon dynamic concepts of stability of motion, it develops asymptotic buckling and postbuckling analyses from potential energy considerations, with applications to columns, plates, and arches. 1974 ed. 208pp. 5⅜ x 8½. 42541-X

METAL FATIGUE, N. E. Frost, K. J. Marsh, and L. P. Pook. Definitive, clearly written, and well-illustrated volume addresses all aspects of the subject, from the historical development of understanding metal fatigue to vital concepts of the cyclic stress that causes a crack to grow. Includes 7 appendixes. 544pp. 5⅜ x 8½. 40927-9

Mathematics

FUNCTIONAL ANALYSIS (Second Corrected Edition), George Bachman and Lawrence Narici. Excellent treatment of subject geared toward students with background in linear algebra, advanced calculus, physics, and engineering. Text covers introduction to inner-product spaces, normed, metric spaces, and topological spaces; complete orthonormal sets, the Hahn-Banach Theorem and its consequences, and many other related subjects. 1966 ed. 544pp. 6⅛ x 9¼. 40251-7

ASYMPTOTIC EXPANSIONS OF INTEGRALS, Norman Bleistein & Richard A. Handelsman. Best introduction to important field with applications in a variety of scientific disciplines. New preface. Problems. Diagrams. Tables. Bibliography. Index. 448pp. 5⅜ x 8½. 65082-0

VECTOR AND TENSOR ANALYSIS WITH APPLICATIONS, A. I. Borisenko and I. E. Tarapov. Concise introduction. Worked-out problems, solutions, exercises. 257pp. 5⅝ x 8¼. 63833-2

THE ABSOLUTE DIFFERENTIAL CALCULUS (CALCULUS OF TENSORS), Tullio Levi-Civita. Great 20th-century mathematician's classic work on material necessary for mathematical grasp of theory of relativity. 452pp. 5⅝ x 8¼. 63401-9

AN INTRODUCTION TO ORDINARY DIFFERENTIAL EQUATIONS, Earl A. Coddington. A thorough and systematic first course in elementary differential equations for undergraduates in mathematics and science, with many exercises and problems (with answers). Index. 304pp. 5⅜ x 8½. 65942-9

FOURIER SERIES AND ORTHOGONAL FUNCTIONS, Harry F. Davis. An incisive text combining theory and practical example to introduce Fourier series, orthogonal functions and applications of the Fourier method to boundary-value problems. 570 exercises. Answers and notes. 416pp. 5⅜ x 8½. 65973-9

COMPUTABILITY AND UNSOLVABILITY, Martin Davis. Classic graduate-level introduction to theory of computability, usually referred to as theory of recurrent functions. New preface and appendix. 288pp. 5⅜ x 8½. 61471-9

ASYMPTOTIC METHODS IN ANALYSIS, N. G. de Bruijn. An inexpensive, comprehensive guide to asymptotic methods–the pioneering work that teaches by explaining worked examples in detail. Index. 224pp. 5⅜ x 8½ 64221-6

APPLIED COMPLEX VARIABLES, John W. Dettman. Step-by-step coverage of fundamentals of analytic function theory–plus lucid exposition of five important applications: Potential Theory; Ordinary Differential Equations; Fourier Transforms; Laplace Transforms; Asymptotic Expansions. 66 figures. Exercises at chapter ends. 512pp. 5⅜ x 8½. 64670-X

INTRODUCTION TO LINEAR ALGEBRA AND DIFFERENTIAL EQUATIONS, John W. Dettman. Excellent text covers complex numbers, determinants, orthonormal bases, Laplace transforms, much more. Exercises with solutions. Undergraduate level. 416pp. 5⅜ x 8½. 65191-6

CALCULUS OF VARIATIONS WITH APPLICATIONS, George M. Ewing. Applications-oriented introduction to variational theory develops insight and promotes understanding of specialized books, research papers. Suitable for advanced undergraduate/graduate students as primary, supplementary text. 352pp. 5⅜ x 8½. 64856-7

COMPLEX VARIABLES, Francis J. Flanigan. Unusual approach, delaying complex algebra till harmonic functions have been analyzed from real variable viewpoint. Includes problems with answers. 364pp. 5⅜ x 8½. 61388-7

AN INTRODUCTION TO THE CALCULUS OF VARIATIONS, Charles Fox. Graduate-level text covers variations of an integral, isoperimetrical problems, least action, special relativity, approximations, more. References. 279pp. 5⅜ x 8½. 65499-0

COUNTEREXAMPLES IN ANALYSIS, Bernard R. Gelbaum and John M. H. Olmsted. These counterexamples deal mostly with the part of analysis known as "real variables." The first half covers the real number system, and the second half encompasses higher dimensions. 1962 edition. xxiv+198pp. 5⅜ x 8½. 42875-3

CATASTROPHE THEORY FOR SCIENTISTS AND ENGINEERS, Robert Gilmore. Advanced-level treatment describes mathematics of theory grounded in the work of Poincaré, R. Thom, other mathematicians. Also important applications to problems in mathematics, physics, chemistry, and engineering. 1981 edition. References. 28 tables. 397 black-and-white illustrations. xvii+666pp. 6⅛ x 9¼. 67539-4

INTRODUCTION TO DIFFERENCE EQUATIONS, Samuel Goldberg. Exceptionally clear exposition of important discipline with applications to sociology, psychology, economics. Many illustrative examples; over 250 problems. 260pp. 5⅜ x 8½. 65084-7

NUMERICAL METHODS FOR SCIENTISTS AND ENGINEERS, Richard Hamming. Classic text stresses frequency approach in coverage of algorithms, polynomial approximation, Fourier approximation, exponential approximation, other topics. Revised and enlarged 2nd edition. 721pp. 5⅜ x 8½. 65241-6

INTRODUCTION TO NUMERICAL ANALYSIS (2nd Edition), F. B. Hildebrand. Classic, fundamental treatment covers computation, approximation, interpolation, numerical differentiation and integration, other topics. 150 new problems. 669pp. 5⅜ x 8½. 65363-3

THREE PEARLS OF NUMBER THEORY, A. Y. Khinchin. Three compelling puzzles require proof of a basic law governing the world of numbers. Challenges concern van der Waerden's theorem, the Landau-Schnirelmann hypothesis and Mann's theorem, and a solution to Waring's problem. Solutions included. 64pp. 5⅜ x 8½. 40026-3

THE PHILOSOPHY OF MATHEMATICS: An Introductory Essay, Stephan Körner. Surveys the views of Plato, Aristotle, Leibniz & Kant concerning propositions and theories of applied and pure mathematics. Introduction. Two appendices. Index. 198pp. 5⅜ x 8½. 25048-2

TENSOR CALCULUS, J.L. Synge and A. Schild. Widely used introductory text covers spaces and tensors, basic operations in Riemannian space, non-Riemannian spaces, etc. 324pp. 5⅜ x 8¼. 63612-7

ORDINARY DIFFERENTIAL EQUATIONS, Morris Tenenbaum and Harry Pollard. Exhaustive survey of ordinary differential equations for undergraduates in mathematics, engineering, science. Thorough analysis of theorems. Diagrams. Bibliography. Index. 818pp. 5⅜ x 8½. 64940-7

INTEGRAL EQUATIONS, F. G. Tricomi. Authoritative, well-written treatment of extremely useful mathematical tool with wide applications. Volterra Equations, Fredholm Equations, much more. Advanced undergraduate to graduate level. Exercises. Bibliography. 238pp. 5⅜ x 8½. 64828-1

FOURIER SERIES, Georgi P. Tolstov. Translated by Richard A. Silverman. A valuable addition to the literature on the subject, moving clearly from subject to subject and theorem to theorem. 107 problems, answers. 336pp. 5⅜ x 8½. 63317-9

INTRODUCTION TO MATHEMATICAL THINKING, Friedrich Waismann. Examinations of arithmetic, geometry, and theory of integers; rational and natural numbers; complete induction; limit and point of accumulation; remarkable curves; complex and hypercomplex numbers, more. 1959 ed. 27 figures. xii+260pp. 5⅜ x 8½. 42804-4

POPULAR LECTURES ON MATHEMATICAL LOGIC, Hao Wang. Noted logician's lucid treatment of historical developments, set theory, model theory, recursion theory and constructivism, proof theory, more. 3 appendixes. Bibliography. 1981 ed. ix+283pp. 5⅜ x 8½. 67632-3

CALCULUS OF VARIATIONS, Robert Weinstock. Basic introduction covering isoperimetric problems, theory of elasticity, quantum mechanics, electrostatics, etc. Exercises throughout. 326pp. 5⅜ x 8½. 63069-2

THE CONTINUUM: A Critical Examination of the Foundation of Analysis, Hermann Weyl. Classic of 20th-century foundational research deals with the conceptual problem posed by the continuum. 156pp. 5⅜ x 8½. 67982-9

CHALLENGING MATHEMATICAL PROBLEMS WITH ELEMENTARY SOLUTIONS, A. M. Yaglom and I. M. Yaglom. Over 170 challenging problems on probability theory, combinatorial analysis, points and lines, topology, convex polygons, many other topics. Solutions. Total of 445pp. 5⅜ x 8½. Two-vol. set.
Vol. I: 65536-9 Vol. II: 65537-7

INTRODUCTION TO PARTIAL DIFFERENTIAL EQUATIONS WITH APPLICATIONS, E. C. Zachmanoglou and Dale W. Thoe. Essentials of partial differential equations applied to common problems in engineering and the physical sciences. Problems and answers. 416pp. 5⅜ x 8½. 65251-3

THE THEORY OF GROUPS, Hans J. Zassenhaus. Well-written graduate-level text acquaints reader with group-theoretic methods and demonstrates their usefulness in mathematics. Axioms, the calculus of complexes, homomorphic mapping, *p*-group theory, more. 276pp. 5⅜ x 8½. 40922-8

Math–Decision Theory, Statistics, Probability

ELEMENTARY DECISION THEORY, Herman Chernoff and Lincoln E. Moses. Clear introduction to statistics and statistical theory covers data processing, probability and random variables, testing hypotheses, much more. Exercises. 364pp. 5⅜ x 8½. 65218-1

STATISTICS MANUAL, Edwin L. Crow et al. Comprehensive, practical collection of classical and modern methods prepared by U.S. Naval Ordnance Test Station. Stress on use. Basics of statistics assumed. 288pp. 5⅜ x 8½. 60599-X

SOME THEORY OF SAMPLING, William Edwards Deming. Analysis of the problems, theory, and design of sampling techniques for social scientists, industrial managers, and others who find statistics important at work. 61 tables. 90 figures. xvii +602pp. 5⅜ x 8½. 64684-X

LINEAR PROGRAMMING AND ECONOMIC ANALYSIS, Robert Dorfman, Paul A. Samuelson and Robert M. Solow. First comprehensive treatment of linear programming in standard economic analysis. Game theory, modern welfare economics, Leontief input-output, more. 525pp. 5⅜ x 8½. 65491-5

PROBABILITY: An Introduction, Samuel Goldberg. Excellent basic text covers set theory, probability theory for finite sample spaces, binomial theorem, much more. 360 problems. Bibliographies. 322pp. 5⅜ x 8½. 65252-1

GAMES AND DECISIONS: Introduction and Critical Survey, R. Duncan Luce and Howard Raiffa. Superb nontechnical introduction to game theory, primarily applied to social sciences. Utility theory, zero-sum games, n-person games, decision-making, much more. Bibliography. 509pp. 5⅜ x 8½. 65943-7

INTRODUCTION TO THE THEORY OF GAMES, J. C. C. McKinsey. This comprehensive overview of the mathematical theory of games illustrates applications to situations involving conflicts of interest, including economic, social, political, and military contexts. Appropriate for advanced undergraduate and graduate courses; advanced calculus a prerequisite. 1952 ed. x+372pp. 5⅜ x 8½. 42811-7

FIFTY CHALLENGING PROBLEMS IN PROBABILITY WITH SOLUTIONS, Frederick Mosteller. Remarkable puzzlers, graded in difficulty, illustrate elementary and advanced aspects of probability. Detailed solutions. 88pp. 5⅜ x 8½. 65355-2

PROBABILITY THEORY: A Concise Course, Y. A. Rozanov. Highly readable, self-contained introduction covers combination of events, dependent events, Bernoulli trials, etc. 148pp. 5⅜ x 8¼. 63544-9

STATISTICAL METHOD FROM THE VIEWPOINT OF QUALITY CONTROL, Walter A. Shewhart. Important text explains regulation of variables, uses of statistical control to achieve quality control in industry, agriculture, other areas. 192pp. 5⅜ x 8½. 65232-7

Math–Geometry and Topology

ELEMENTARY CONCEPTS OF TOPOLOGY, Paul Alexandroff. Elegant, intuitive approach to topology from set-theoretic topology to Betti groups; how concepts of topology are useful in math and physics. 25 figures. 57pp. 5⅜ x 8½. 60747-X

COMBINATORIAL TOPOLOGY, P. S. Alexandrov. Clearly written, well-organized, three-part text begins by dealing with certain classic problems without using the formal techniques of homology theory and advances to the central concept, the Betti groups. Numerous detailed examples. 654pp. 5⅜ x 8½. 40179-0

EXPERIMENTS IN TOPOLOGY, Stephen Barr. Classic, lively explanation of one of the byways of mathematics. Klein bottles, Moebius strips, projective planes, map coloring, problem of the Koenigsberg bridges, much more, described with clarity and wit. 43 figures. 210pp. 5⅜ x 8½. 25933-1

CONFORMAL MAPPING ON RIEMANN SURFACES, Harvey Cohn. Lucid, insightful book presents ideal coverage of subject. 334 exercises make book perfect for self-study. 55 figures. 352pp. 5⅝ x 8¼. 64025-6

THE GEOMETRY OF RENÉ DESCARTES, René Descartes. The great work founded analytical geometry. Original French text, Descartes's own diagrams, together with definitive Smith-Latham translation. 244pp. 5⅜ x 8½. 60068-8

PRACTICAL CONIC SECTIONS: The Geometric Properties of Ellipses, Parabolas and Hyperbolas, J. W. Downs. This text shows how to create ellipses, parabolas, and hyperbolas. It also presents historical background on their ancient origins and describes the reflective properties and roles of curves in design applications. 1993 ed. 98 figures. xii+100pp. 6½ x 9¼. 42876-1

THE THIRTEEN BOOKS OF EUCLID'S ELEMENTS, translated with introduction and commentary by Thomas L. Heath. Definitive edition. Textual and linguistic notes, mathematical analysis. 2,500 years of critical commentary. Unabridged. 1,414pp. 5⅜ x 8½. Three-vol. set. Vol. I: 60088-2 Vol. II: 60089-0 Vol. III: 60090-4

GEOMETRY OF COMPLEX NUMBERS, Hans Schwerdtfeger. Illuminating, widely praised book on analytic geometry of circles, the Moebius transformation, and two-dimensional non-Euclidean geometries. 200pp. 5⅝ x 8¼. 63830-8

DIFFERENTIAL GEOMETRY, Heinrich W. Guggenheimer. Local differential geometry as an application of advanced calculus and linear algebra. Curvature, transformation groups, surfaces, more. Exercises. 62 figures. 378pp. 5⅜ x 8½. 63433-7

CURVATURE AND HOMOLOGY: Enlarged Edition, Samuel I. Goldberg. Revised edition examines topology of differentiable manifolds; curvature, homology of Riemannian manifolds; compact Lie groups; complex manifolds; curvature, homology of Kaehler manifolds. New Preface. Four new appendixes. 416pp. 5⅜ x 8½. 40207-X

History of Math

THE WORKS OF ARCHIMEDES, Archimedes (T. L. Heath, ed.). Topics include the famous problems of the ratio of the areas of a cylinder and an inscribed sphere; the measurement of a circle; the properties of conoids, spheroids, and spirals; and the quadrature of the parabola. Informative introduction. clxxxvi+326pp; supplement, 52pp. 5⅜ x 8½. 42084-1

A SHORT ACCOUNT OF THE HISTORY OF MATHEMATICS, W. W. Rouse Ball. One of clearest, most authoritative surveys from the Egyptians and Phoenicians through 19th-century figures such as Grassman, Galois, Riemann. Fourth edition. 522pp. 5⅜ x 8½. 20630-0

THE HISTORY OF THE CALCULUS AND ITS CONCEPTUAL DEVELOPMENT, Carl B. Boyer. Origins in antiquity, medieval contributions, work of Newton, Leibniz, rigorous formulation. Treatment is verbal. 346pp. 5⅜ x 8½. 60509-4

THE HISTORICAL ROOTS OF ELEMENTARY MATHEMATICS, Lucas N. H. Bunt, Phillip S. Jones, and Jack D. Bedient. Fundamental underpinnings of modern arithmetic, algebra, geometry, and number systems derived from ancient civilizations. 320pp. 5⅜ x 8½. 25563-8

A HISTORY OF MATHEMATICAL NOTATIONS, Florian Cajori. This classic study notes the first appearance of a mathematical symbol and its origin, the competition it encountered, its spread among writers in different countries, its rise to popularity, its eventual decline or ultimate survival. Original 1929 two-volume edition presented here in one volume. xxviii+820pp. 5⅜ x 8½. 67766-4

GAMES, GODS & GAMBLING: A History of Probability and Statistical Ideas, F. N. David. Episodes from the lives of Galileo, Fermat, Pascal, and others illustrate this fascinating account of the roots of mathematics. Features thought-provoking references to classics, archaeology, biography, poetry. 1962 edition. 304pp. 5⅜ x 8½. (Available in U.S. only.) 40023-9

OF MEN AND NUMBERS: The Story of the Great Mathematicians, Jane Muir. Fascinating accounts of the lives and accomplishments of history's greatest mathematical minds–Pythagoras, Descartes, Euler, Pascal, Cantor, many more. Anecdotal, illuminating. 30 diagrams. Bibliography. 256pp. 5⅜ x 8½. 28973-7

HISTORY OF MATHEMATICS, David E. Smith. Nontechnical survey from ancient Greece and Orient to late 19th century; evolution of arithmetic, geometry, trigonometry, calculating devices, algebra, the calculus. 362 illustrations. 1,355pp. 5⅜ x 8½. Two-vol. set. Vol. I: 20429-4 Vol. II: 20430-8

A CONCISE HISTORY OF MATHEMATICS, Dirk J. Struik. The best brief history of mathematics. Stresses origins and covers every major figure from ancient Near East to 19th century. 41 illustrations. 195pp. 5⅜ x 8½. 60255-9

Physics

OPTICAL RESONANCE AND TWO-LEVEL ATOMS, L. Allen and J. H. Eberly. Clear, comprehensive introduction to basic principles behind all quantum optical resonance phenomena. 53 illustrations. Preface. Index. 256pp. 5⅜ x 8½. 65533-4

QUANTUM THEORY, David Bohm. This advanced undergraduate-level text presents the quantum theory in terms of qualitative and imaginative concepts, followed by specific applications worked out in mathematical detail. Preface. Index. 655pp. 5⅜ x 8½. 65969-0

ATOMIC PHYSICS: 8th edition, Max Born. Nobel laureate's lucid treatment of kinetic theory of gases, elementary particles, nuclear atom, wave-corpuscles, atomic structure and spectral lines, much more. Over 40 appendices, bibliography. 495pp. 5⅜ x 8½. 65984-4

A SOPHISTICATE'S PRIMER OF RELATIVITY, P. W. Bridgman. Geared toward readers already acquainted with special relativity, this book transcends the view of theory as a working tool to answer natural questions: What is a frame of reference? What is a "law of nature"? What is the role of the "observer"? Extensive treatment, written in terms accessible to those without a scientific background. 1983 ed. xlviii+172pp. 5⅜ x 8½. 42549-5

AN INTRODUCTION TO HAMILTONIAN OPTICS, H. A. Buchdahl. Detailed account of the Hamiltonian treatment of aberration theory in geometrical optics. Many classes of optical systems defined in terms of the symmetries they possess. Problems with detailed solutions. 1970 edition. xv+360pp. 5⅜ x 8½. 67597-1

PRIMER OF QUANTUM MECHANICS, Marvin Chester. Introductory text examines the classical quantum bead on a track: its state and representations; operator eigenvalues; harmonic oscillator and bound bead in a symmetric force field; and bead in a spherical shell. Other topics include spin, matrices, and the structure of quantum mechanics; the simplest atom; indistinguishable particles; and stationary-state perturbation theory. 1992 ed. xiv+314pp. 6⅛ x 9¼. 42878-8

LECTURES ON QUANTUM MECHANICS, Paul A. M. Dirac. Four concise, brilliant lectures on mathematical methods in quantum mechanics from Nobel Prize–winning quantum pioneer build on idea of visualizing quantum theory through the use of classical mechanics. 96pp. 5⅜ x 8½. 41713-1

THIRTY YEARS THAT SHOOK PHYSICS: The Story of Quantum Theory, George Gamow. Lucid, accessible introduction to influential theory of energy and matter. Careful explanations of Dirac's anti-particles, Bohr's model of the atom, much more. 12 plates. Numerous drawings. 240pp. 5⅜ x 8½. 24895-X

ELECTRONIC STRUCTURE AND THE PROPERTIES OF SOLIDS: The Physics of the Chemical Bond, Walter A. Harrison. Innovative text offers basic understanding of the electronic structure of covalent and ionic solids, simple metals, transition metals and their compounds. Problems. 1980 edition. 582pp. 6⅛ x 9¼. 66021-4